W9-CNA-570

Euclid Public Library
631 E. 222nd Street
Euclid, Ohio 44123
216-261-5300

33 Men

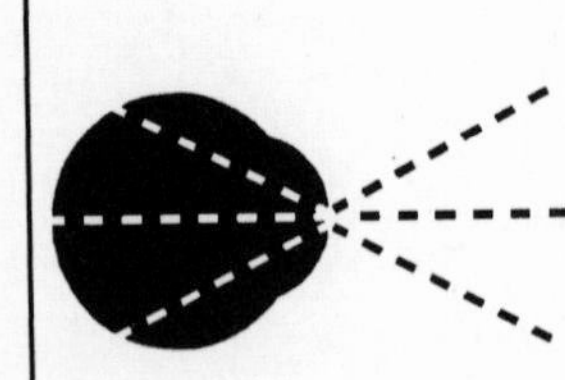

This Large Print Book carries the Seal of Approval of N.A.V.H.

33 Men

INSIDE THE MIRACULOUS SURVIVAL AND DRAMATIC RESCUE OF THE CHILEAN MINERS

JONATHAN FRANKLIN

THORNDIKE PRESS
A part of Gale, Cengage Learning

Detroit • New York • San Francisco • New Haven, Conn • Waterville, Maine • London

GALE
CENGAGE Learning™

Copyright © 2011 by Jonathan Franklin.

Map and Diagram by Jeffrey L. Ward.

Thorndike Press, a part of Gale, Cengage Learning.

ALL RIGHTS RESERVED

While the author has made every effort to provide accurate telephone numbers and Internet addresses at the time of publication, neither the publisher nor the author assumes any responsibility for errors, or for changes that occur after publication. Further, the publisher does not have any control over and does not assume any responsibility for author or third-party websites or their content.

Thorndike Press® Large Print Nonfiction.

The text of this Large Print edition is unabridged.

Other aspects of the book may vary from the original edition.

Set in 16 pt. Plantin.

LIBRARY OF CONGRESS CATALOGING-IN-PUBLICATION DATA

Franklin, Jonathan, 1964–
33 men : inside the miraculous survival and dramatic rescue of the Chilean miners / by Jonathan Franklin. — Large print ed.
p. cm.
ISBN-13: 978-1-4104-3662-7 (hardcover : large print)
ISBN-10: 1-4104-3662-4 (hardcover : large print) 1. Gold mines and mining — Accidents — Chile — Copiapó Region. 2. Copper mines and mining — Accidents — Chile — Copiapó Region. 3. Mine rescue work — Chile —Copiapó Region. 4. Survival after airplane accidents, shipwrecks, etc. — Chile — Copiapó Region. 5. Gold miners — Chile — Biography. 6. Copper miners — Chile — Biography. 7. Copiapó Region (Chile) — History — 21st century. 8. Copiapó Region (Chile) — Biography. 9. Large type books. I. Title. II. Title: Thirty-three men.
TN311.F73 2011b
363.11'96223430983145—dc22

2010053745

Published in 2011 by arrangement with G. P. Putnam's Sons, a member of Penguin Group (USA) Inc.

Printed in the United States of America
1 2 3 4 5 6 7 15 14 13 12 11

This book is dedicated to my family, who barely saw me for the duration of this dramatic tale: Toty, my ever patient and daredevil wife, and my six precious daughters, Francisca, Susan, Maciel, Kimberly, Amy and little Zoe. And finally to my grandson Tomas, who hardly saw me.

Writing this book was a challenge and a journey, not nearly as wrenching as that lived by the thirty-three miners, but I, too, am excited to finally be at home and at peace.

JONATHAN FRANKLIN
December 2010
Santiago, Chile

CONTENTS

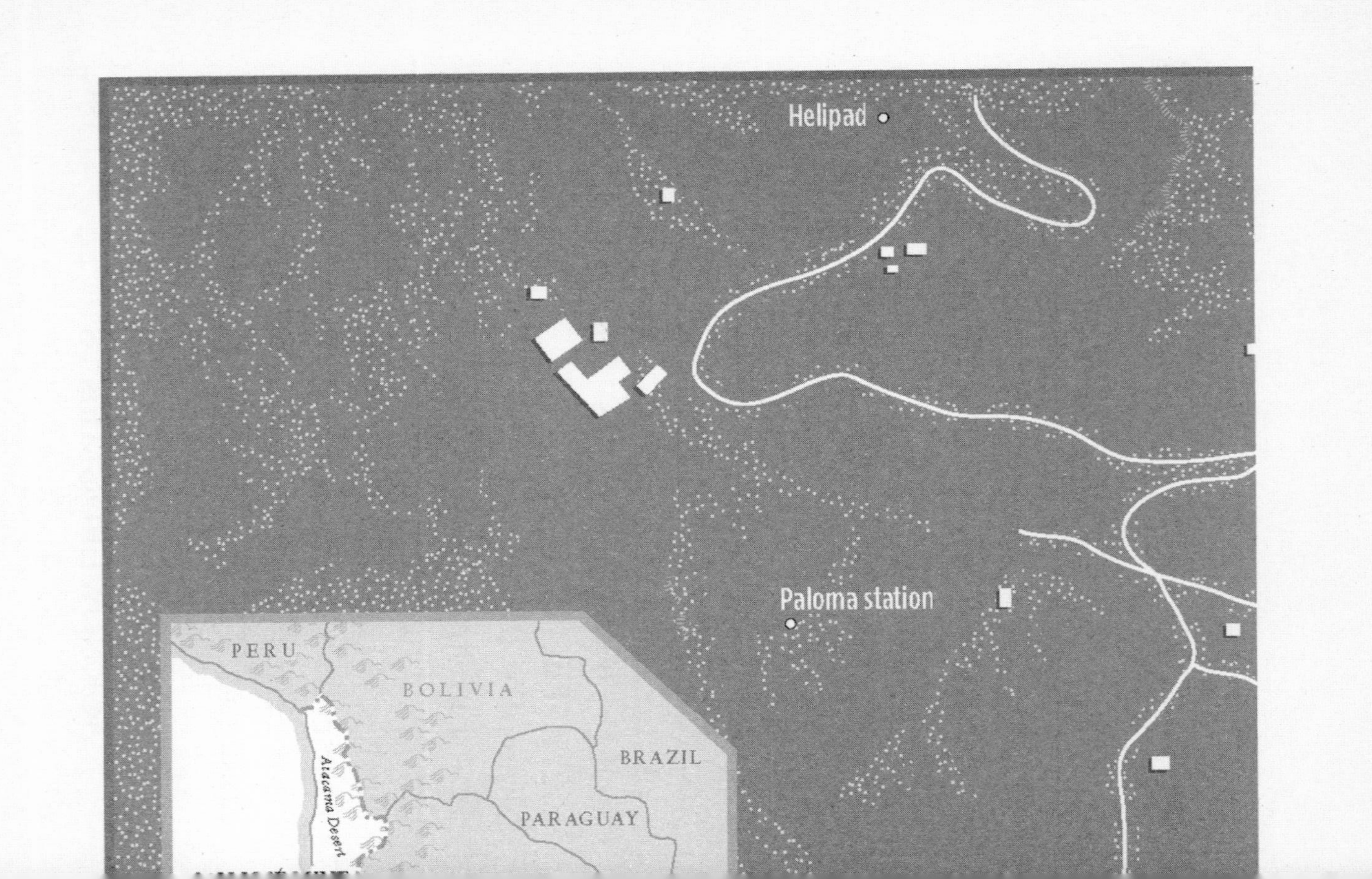
Helipad
Paloma station
PERU
BOLIVIA
BRAZIL
PARAGUAY
Atacama Desert

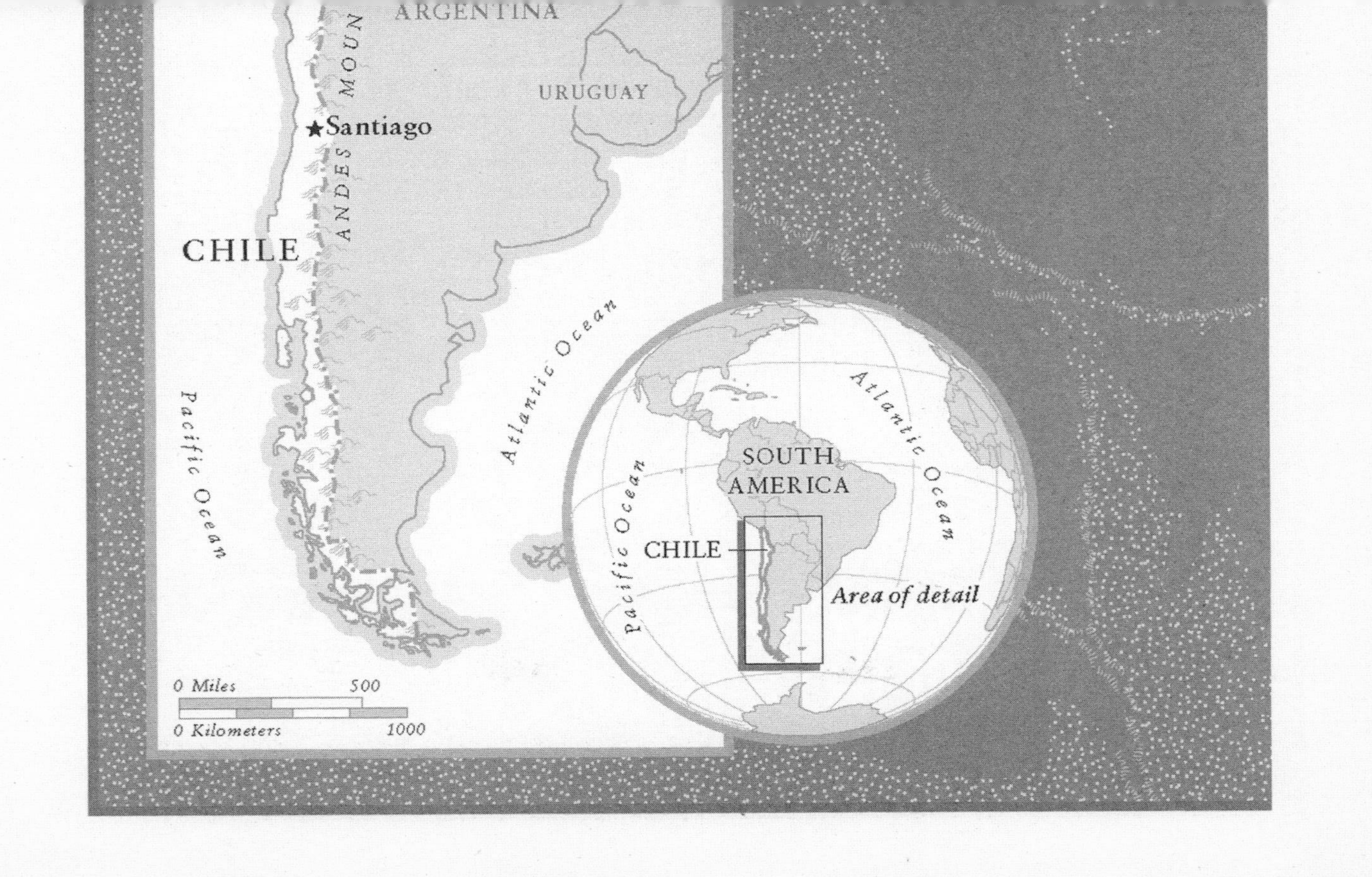
ARGENTINA
ANDES MOUN
URUGUAY
Santiago
CHILE
Pacific Ocean
Atlantic Ocean
SOUTH AMERICA
CHILE
Pacific Ocean
Atlantic Ocean
Area of detail
0 Miles 500
0 Kilometers 1000

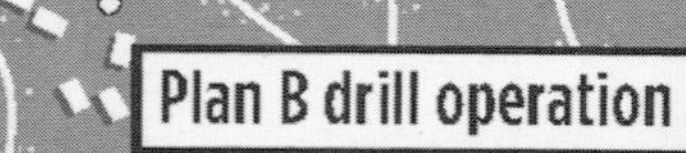

SAN JOSÉ MINE
RESCUE OPERATION
N
Plan C drill operation
Plan B drill operation
Field hospital
Drill workshop

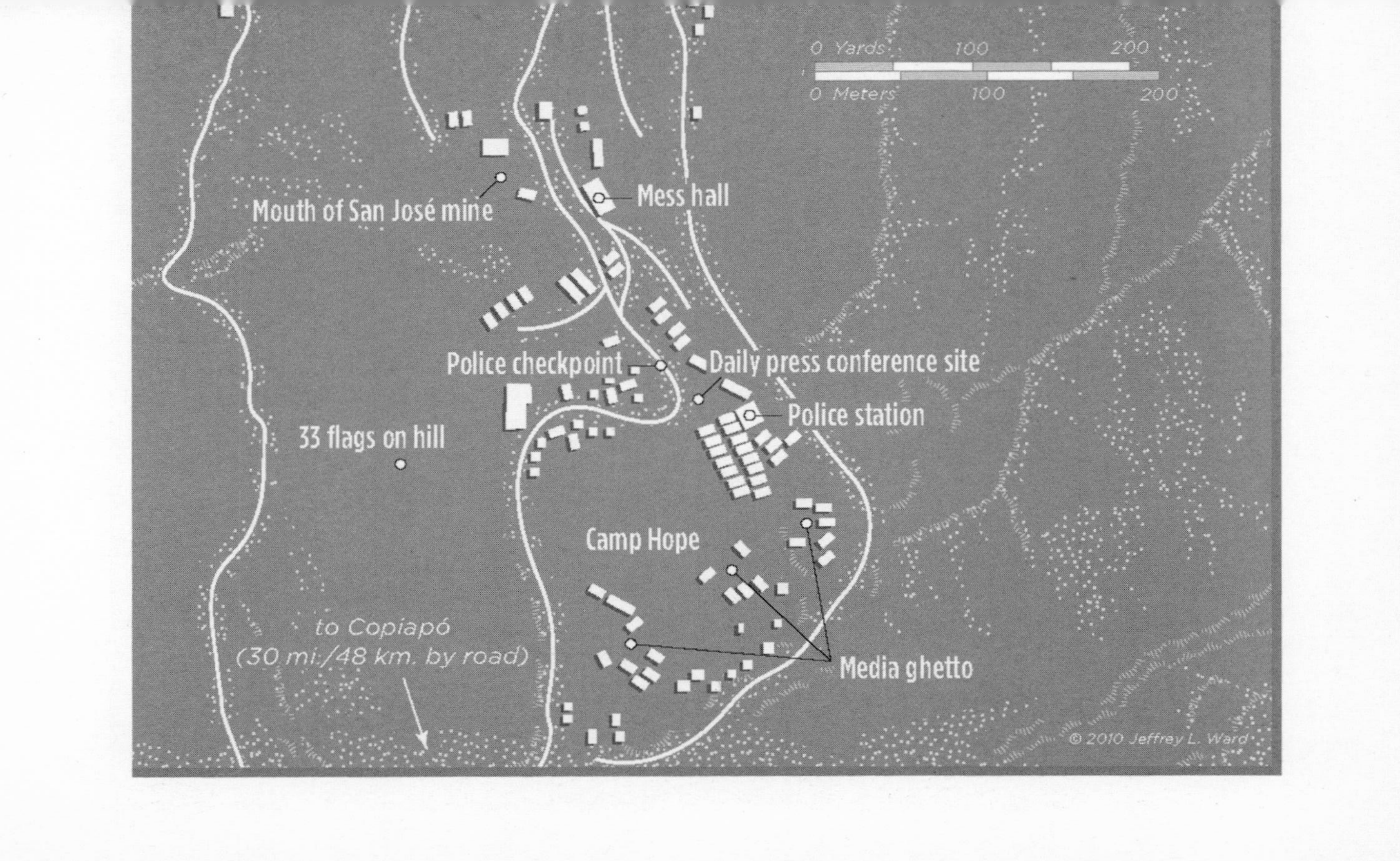
0 Yards 100 200
0 Meters 100 200
Mouth of San José mine
Mess hall
Police checkpoint
Daily press conference site
Police station
33 flags on hill
Camp Hope
Media ghetto
to Copiapó
(30 mi./48 km. by road)
© 2010 Jeffrey L. Ward

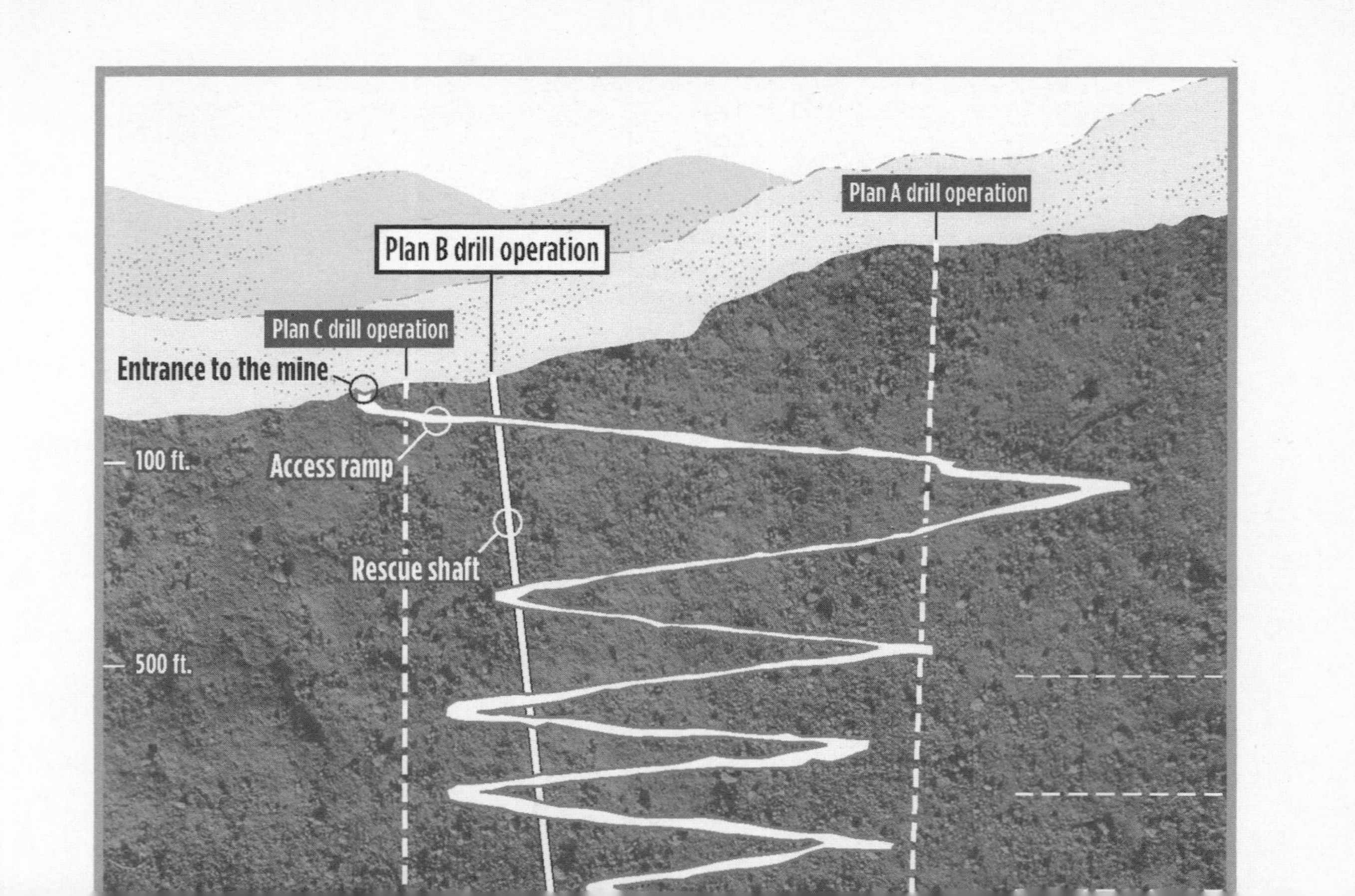
Plan A drill operation
Plan B drill operation
Plan C drill operation
Entrance to the mine
Access ramp
Rescue shaft
100 ft.
500 ft.

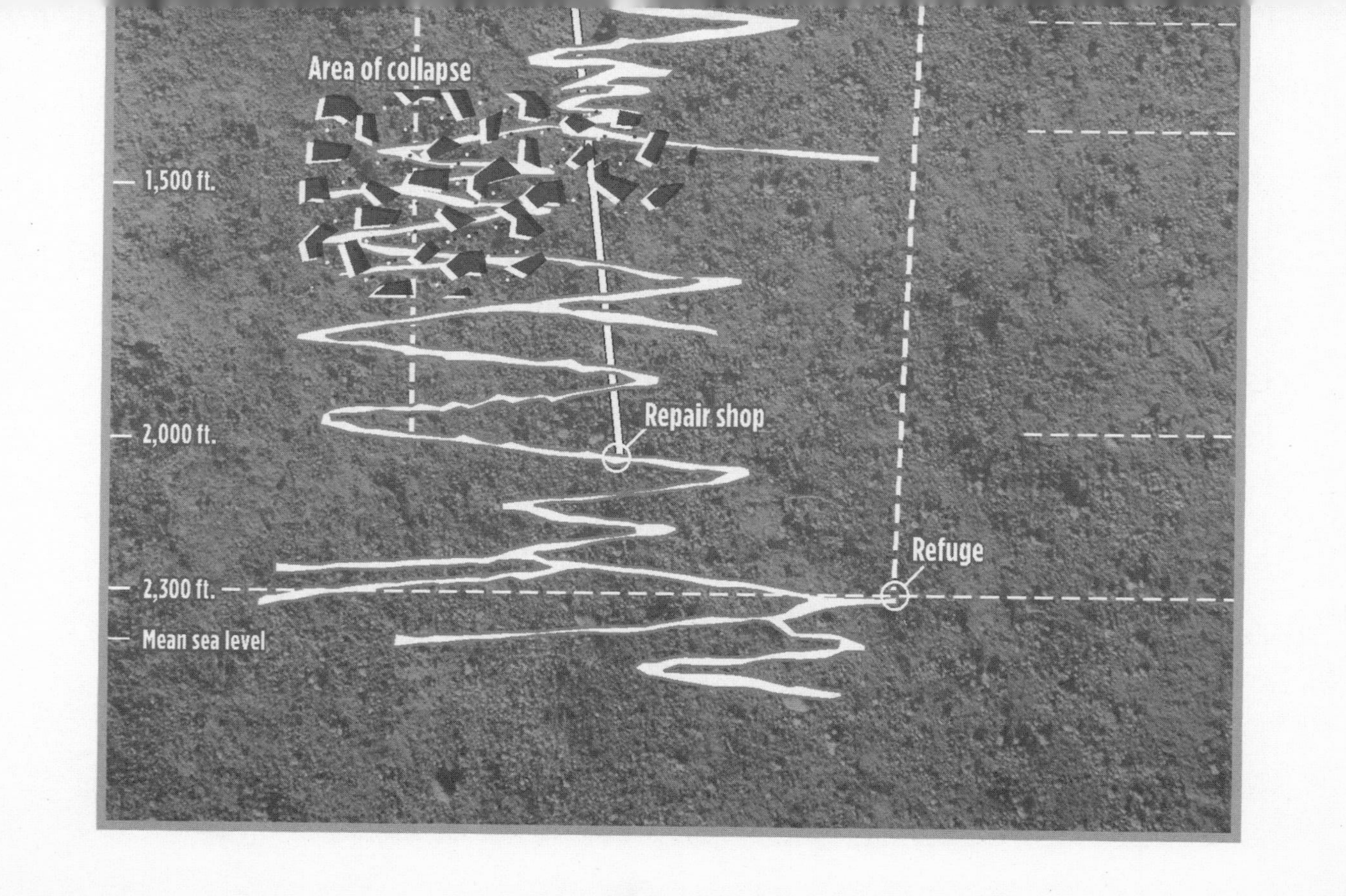
Area of collapse
1,500 ft.
2,000 ft.
Repair shop
Refuge
2,300 ft.
Mean sea level

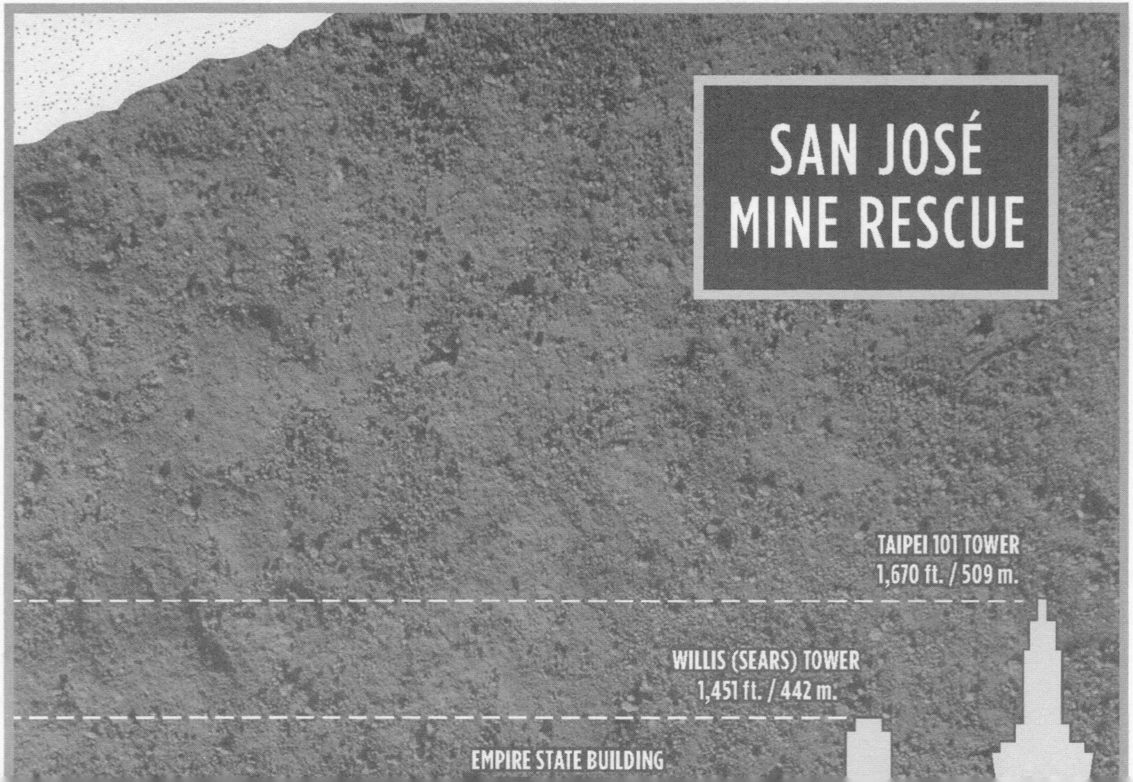

SAN JOSÉ MINE RESCUE

TITANIC
882 ft. /269 m.
STATUE OF LIBERTY
305 ft. / 93 m.
Depth reference
© 2010 Jeffrey L. Ward

Prologue The Eyes of the World

On October 12, a dense dawn fog covered a packed mountainside in northern Chile. Dreamy banks of mist climbed up the slopes. The sun was still hidden over the horizon; a cold damp air rose up from the Pacific Ocean and sucked away body heat. The few figures that meandered through the makeshift camp at this early hour were ghostly silhouettes — like fleeting mirages here in the Atacama Desert, one of the world's driest locations.

In the media encampment, a maze of floodlights illuminated fields of antennae. Dozens of satellite transmitters were propped atop a field of boulders.

Huddled around a campfire, fingers and arms entwined, the Ávalos family prayed and talked in quiet reverence directly above two buried relatives: twenty-nine-year-old Renán and thirty-one-year-old

Florencio Ávalos. Nine weeks earlier, on August 5, the brothers had entered the San José mine for a twelve-hour shift. By mid-afternoon a massive slab of rock — the size of a skyscraper — had sheared off the mountain and sealed them at the bottom of the mine.

For nine weeks the Ávalos family had hoped and prayed for a miracle. First to hear word that the brothers were alive and then to have them safely rescued from the depths of a mine that even on the best days was notorious for killing and maiming miners.

From the moment the mine collapsed in early August, hundreds of professional engineers, rescue workers, drillers and diggers had descended on this previously remote and deserted corner of northern Chile. They arrived to volunteer, offering their ideas, their equipment and their hard labor. Using both diplomatic channels and contacts in the business community, Chilean President Sebastián Piñera made a simple but profound call for help. He said, "We have these guys trapped at seven hundred meters [2,300 feet]. What technologies do you have that could possibly help?"

The response was overwhelming.

Now the rescue was in its final stage. In less than twenty-four hours, a rocket-shaped capsule known as the Phoenix would be slowly lowered into the earth, to the bottom of the mine. Florencio Ávalos would be the first miner to open the door of the contraption and attempt to ride to the surface. His family knew the designation was both an honor and a risk.

Hundreds of rescue workers had worked for months for this moment, most of them laboring in silence. All now brimmed with pride at the chance to play a small part in what was increasingly a global drama and what they all knew was a massive experiment. Never before had miners been rescued from such a depth after months of entrapment. Despite numerous theories that such a rescue was possible, everyone knew the law of averages — never great in an industry as dangerous as mining — was stacked against the probability that all the men could be rescued alive.

Named "Operation San Lorenzo" — in homage to Saint Lorenzo, the patron saint of miners — the rescue was led by Codelco, the Chilean state-owned mining company that, over the past two months, had gathered the world's most sophisticated drilling and mapping equipment.

Codelco, a modern company with profits in excess of $4.5 billion a year, had used a fleet of borrowed, rented and improvised drilling rigs to find the men, and to feed them for sixty-nine days. Now was the moment of truth. Could they pull the men to safety? From a depth more than twice as high as the Eiffel Tower? The rescue hole was so small that the miners had been instructed to exercise strenuously to make sure they would actually fit inside the capsule.

Despite the early hour, hundreds of journalists were already awake, lugging camera equipment in an effort to reserve a privileged spot for a drama that had captured the hearts and imaginations of viewers worldwide. Not since the moon landing had a technical challenge so intrigued and captivated the world. And in 2010, the wired world offered dozens of new ways to follow and comment on the proceedings.

With their heads bowed toward a scorching mound of orange embers, testament to weeks of waiting, the Ávalos family appeared oblivious to the growing commotion. They offered a few comments, then ignored the arrival of a stray cameraman. The journalist — with cables and soundman in tow — waded in for a few minutes of live broadcasting, every word transmitted to an audience around the world, and then migrated to the next family.

Behind the Ávalos family was the banner "Buried Perhaps . . . Overcome Never." The miners' faces stared out from the sign, half hidden in the dark. As individuals, their faces were unremarkable — serious, dour and weathered. As a group, they were Los 33, a worldwide symbol of resilience.

Throughout September and October 2010, as the rescuers drilled through a mountain of granite in search of the trapped men, the fate of Los 33 had become a collective narrative. The world's leading journalists flooded in, battling over scarce airline tickets to travel to Copiapó, a city so forgotten that when Chilean broadcasters delivered the national weather forecast, this was the only major city in Chile they simply skipped. "When the World Cup trophy toured all of Chile, they didn't stop here," groused Copiapó mayor Maglio Cicardini, a ponytailed showman who looked like a backup rock-and-roll guitarist for ZZ Top.

Despite the worldwide interest, the cameras rarely were allowed access to either the front lines or below the surface of this worldwide tragedy. Locked behind police lines by a strict and slick public relations campaign run by Chilean President Sebastian Piñera, most reporters were reduced to two months of interviewing family members and politicians, while a world audience measured in the hundreds of millions was transfixed by a more profound story line: What was happening down below?

Entombed in a sweltering, humid and crumbling cave, how could thirty-three miners be alive after all these weeks?

By early afternoon, the final countdown had begun. Crowds of relatives stood in awe as huge TV screens mounted on the side of motor homes and on the flaps of the press tent showed images of the rescue workers putting the final touches on the Phoenix rescue capsule. Painted the colors of the Chilean flag — blue, white and red — the Phoenix had been built on specifications developed by NASA and the Chilean Navy.

At 11 PM, the Phoenix was ready. A winch hoisted the capsule. A yellow wheel threaded the cable and slowly rotated. The scene was hypnotic. It looked like an industrial operation from the 1930s. Hidden from view were the modern tools that made the whole operation possible, including GPS units that allowed the massive drills to find a tiny underground target, miles of fiber optic cable and wireless transmitters that relayed the miners' pulse and blood pressure to a physician's laptop.

Sixty-nine days earlier, the men had been lost underground. More than two weeks of searching had failed to find the tunnel where they were slowly starving to death. Death was so certain that the men had written goodbye letters. The government had even begun to design a white cross on the hillside to mark their tomb. Now they might be reborn, resuscitated and rescued. Could such an incredible feat actually be pulled off?

As the world held its breath, the Phoenix was slowly lowered, and then it was gone. In a land of massive earthquakes, the number of ways for the rescue to fail were too numerous to calculate. In order to work, the rescue required not only precise engineering but also a leap of faith. Specialists from around the world had been consulted throughout the rescue, helping to develop medical plans and engineering protocol.

Now even the NASA team was speechless. On this mission, the manual would be written by the Chileans.

One
Buried Alive

Thursday, August 5, 7 AM

The fifty-minute commute to the San José mine was more beautiful than ever. Fields of tiny purple flowers painted the hills in sensual curves, bringing thousands of tourists flocking to view the "flowering desert." Few of the workers aboard the bus noticed; many were asleep as the bus careened around the curves up to the San José mine — a nondescript hill so packed with gold and copper that for more than a century, miners had burrowed like badgers, leaving zigzag tunnels that chased the valuable veins of minerals that laced the mountain innards like blood vessels through a body.

Inside the bus, Mario Gómez could not sleep. The cell phone alarm that had awakened him at 6 AM had been so early and annoying he had asked his wife, "Should I go?" "Skip work," his wife, Lillian, urged. She had long been encouraging her sixty-three-year-old husband to file his retirement papers. Gómez did not need much convincing. He had begun life in the mines when he was twelve years old, a Dickens-like experience, and over the next fifty-one years he had learned every possible permutation of how to die underground. His left hand was a reminder of one way: a dynamite charge had exploded too close and ripped two of his fingers clean off. His thumb was torn away above the knuckle.

From the window of the bus, Gómez watched a desert that offered not a shrub or a tree yet felt full of life compared with the underground world the drowsy men were about to enter. The San José mine was the most dangerous mine in the region and one that not coincidentally paid abnormally high wages. Where else could a *cargador de tiro* — who spent the day cramming sticks of dynamite into recently drilled holes — earn such a succulent salary? The paycheck explained why the men (who called themselves "The Kamikazes") maintained loyalty to their job despite the mine's fearsome reputation. Every worker had come to the same conclusion after completing the cold calculation of danger vs. cash. The cash always won.

As the bus sped along the serpentine road, it passed a row of small altars (*animitas*), each a shrine to a tragic, violent and sudden death. In local lore, an accidental death leaves the dead person's soul in limbo between heaven and earth. By building these shrines, family and friends sought to expedite their loved one's journey skyward, which explains why the lonely temples had lit candles, fresh flowers and crisp photos of the victim. Days later, this same route would have dozens more.

Many of the men carried a hearty lunch. While the mine owners calculated that two sandwiches and a carton of milk were sufficient energy for the twelve-hour shift, the men often brought reinforcements — a bar of chocolate, a Thermos of soup, a neatly wrapped steak and tomato sandwich. And water. Bottles, canteens, even 500cc (2 cups) in plastic bags sold at the Unimarc supermarket. Inside the mine, the temperature rarely dipped below 90 degrees Fahrenheit and the men guzzled three liters (about 3 quarts) of fresh water a day, yet still lived on the fragile border of dehydration. Humidity was so thick that their cigarettes would routinely droop in submission to the elements.

At the entrance to the mine, the men changed into their work clothes: work pants, T-shirt, helmet and head lamp. A simple metal card holder marked their presence —

and often their absence. With seven days of work, then seven days of rest, the men were living the ultimate boom and bust cycle — sweating like animals for a week then whirling in the pleasures of instant excess during the "down week." When the men missed work on Monday they would jokingly say they were paying homage to the god of hangovers, locally known as "Saint Monday."

Company BBQs were frequent, and the owners were known to look the other way when workers showed up hours late. Working on a barren hill, the estimated 250 workers of San Esteban Primera (the holding company for several mines in the region, including San José) had no cell phone coverage, few safety features, frequent accidents, and a near total absence of women. Though it was 2010, in many ways the men lived a frontier existence. The countryside is pockmarked with signs that this is mining country — ranging from the all-night brothels ($40 a shag), to the rows of rugged pickups at Antay, a recently inaugurated casino helping the miners indulge what appeared to be a genetic predisposition to squander a month's salary in a single binge.

The northern deserts of Chile are the world's largest producer of copper, and most Chilean miners work in modern copper mines under the supervision of highly professional multinational companies including Anglo American and BHP Billiton.

With more than 50 percent of the nation's export earnings coming from mining, Chile has long been a world leader in both mining technology and mining operations. Chuquicamata, the world's largest open-pit mine, is run by the Chilean government copper company known as Codelco.

Mining jobs are highly coveted as both lucrative and safe — considering that "safety" in the world of mining is relative. Combine the risks of young men driving truckloads of ammonium nitrate explosives, hundreds of miners setting dynamite charges inside caves every day, and all of this taking place in Chile, a nation known to have the world's biggest earthquakes, and accidents are almost a certainty. Factor in a Chilean party culture fueled by massive quantities of cheap yet head-poundingly strong grape brandy known as *pisco,* and the equation was known to every emergency room nurse in the region: dead miners.

The men entering the San José mine worked not at the safe modern mines but instead belonged to the most risky subculture of this entire industry — low-tech, rustic miners known locally as "Los Pirquineros." While the classic Chilean *pirquinero* had equipment no more sophisticated than a donkey and a pickax, the men at the San José mine called themselves "mechanized *pirquineros,*" meaning they operated modern machinery inside the rickety infrastructure of a classically dangerous operation. Unlike other mines that had rats and insects, the San José mine was sterile — except for the occasional scorpion. Inside the mine, the daily routine was akin to the lifestyle of a California forty-niner searching for gold in the days of Abraham Lincoln. These miners were regularly crushed — "ironed flat" in local lingo — by thousand-pound blocks of rock that unlatched from the roof with terrifying regularity. The rocks inside the San José mine were so sharp that the miners knew that even brushing up against the wall was like scraping a razor across their skin.

A stark reminder of the potential risks came on July 5, 2010. The miners of San José had watched first the rescue operation, then the disappearing pickup that hauled away what was left of Gino Cortés. A block of rock that weighed the equivalent of twenty refrigerators let loose as Gino passed underneath. His leg had been severed clean off. For a moment he looked at his amputated leg in wonder. The cut was so swift he initially felt no pain. A coworker had gingerly brought the leg, wrapped in a shirt, along with Cortés to the emergency room. As he reflected on the accident from his hospital bed in Santiago, Cortés repeated, "I am lucky," as he thanked God for having intact both his right leg and his life. Yet there is no mistaking the crude violence — his now mutilated left leg is sewn neatly into a sausage-like knot below the knee.

If they are not ironed flat, *pirquineros* slowly die from lung problems. Just two months before, miner Alex Vega had been walking in the mine when his legs gave way and he collapsed. Toxic gases from the exhaust of the machinery had stripped his body of oxygen. An ambulance rushed Vega to the local Copiapó hospital, where he was kept for the better part of a week.

Long-term exposure to the gases and grit led to silicosis — caused by breathing toxic silica particles, which clog the lungs. Year after year, these miners inhale clouds of tiny rock fragments, making the lungs ever less efficient. In advanced cases, known as Potter's Rot (in reference to the use of silica in pottery making), the victim lacks oxygen and his skin takes on a blue tint. Mario Gómez, the oldest man on the shift, had collected so much dust and debris in his lungs that after fifty-one years as a miner, he was often short of breath and used a bronchial dilator to maximize the portions of his lungs that still functioned. With silicosis, miners like Gómez are slowly starved of oxygen — essentially the same process that would happen to a pickup truck if it were driven through this desert for twenty years with never a change of air filter.

A *pirquinero* devotes his life to mining for a week, sometimes a solid month, as he breaks his back in solitary battle with the mountain and, for some, then soothes his loneliness with impromptu sexual escapes that a local doctor described as a "*Brokeback Mountain* situation." A Chilean psychiatrist working with these miners described the phenomenon as "transitory homosexuality," which, he noted, is a centuries-old practice among sailors, what he called "a practical solution to the ever more desperate lack of female companionship." When the miners returned to town, they indulged heavily in alcohol, women and a blast of instant pleasures that guaranteed they would soon need another paycheck. Local cocaine — at $15 a gram — was also for many on the list of temptations.

Samuel Ávalos had spent the past twenty-four hours scrambling to earn 16,000 Chilean pesos ($32) to take the bus to Copiapó. Ávalos, a round-faced, hardened man, lived in Rancagua, a mining town just south of Santiago, home to "El Teniente," the world's largest underground mine. Despite the plethora of mining jobs in the area, Ávalos had little experience underground. His job was as a street vendor — his specialty, pirated music CDs. The police harassed him often, sometimes confiscating his stash. But the last day had been lucky — he'd made just enough money to board the last bus with an empty seat to Copiapó. Only later would he realize that José Henríquez, a fellow miner, was on the same bus.

During the bus ride, Ávalos drank. He transferred to the shuttle bus to the mine still in a daze. "The drinks had their effect. Getting down, stepping off the bus, I practically fell," said Ávalos. "Then it was very strange. I don't know what you would call it, but a spirit passed by. My mother. She's deceased. I asked her, 'Mom, what are you saying? What do you want?' I didn't figure it out. Later I had lots of time to think about that last warning."

Ávalos typically stuffed his jacket with chocolates, cakes, cookies, milk, and juice. With his jacket bulging, he constantly battled to hide the contraband from Luis Urzúa, the foreman who was never happy to see his workers with food. He considered it a distraction.

"That day I left my food above. I didn't bring even a single chocolate," said Ávalos. It was another moment he would relive again and again in the coming weeks.

As the incoming shift changed clothes and prepared for work, forty-two-year-old paramedic Hugo Araya exited the mine, his shift complete. Even after six years in San José, Araya never felt comfortable inside the mine. The sagging entrance with that rusted sign about safety always seemed a bit of a joke, considering the constant flow of accidents, cave-ins and fainting miners. But then Araya, who worked as lead emergency medical technician in the mine, was the kind of guy you called in when problems arose. Most of all, he hated the mine's smell. "Like something decomposing. Like rotten meat," he'd say.

With carbon monoxide from the vehicles, gases emanating from the dynamite charges and the men smoking cigarettes nonstop, Araya received the emergency call so often it rarely felt like an emergency anymore. He would then drive the twenty-five-minute, four-mile journey down switchbacks and tunnels, deep to the bottom of the cavern where he'd find a pair of miners sucking on oxygen masks, ready for evacuation. Usually the men could go home that night. At worst, after a day or two in the local clinic they'd be back at the job, hacking, dynamiting, sucking up dust and rarely complaining.

After his full night's shift, Araya was coated with a fine layer of coffee-gray dust, an oily mixture that did not easily wash away. That morning as he showered and scrubbed at his home an hour away in Copiapó, Araya felt a deep unease. The mountain had "cried" all night. Eerie creaking groans and sharp reports had left all the men on edge. When a mine like San José cries, the tears tend to be the size of boulders.

More than a century of picks, dynamite and drills had riddled the mountain with so many holes and tunnels that new workers would wonder aloud how the roof did not fall down on the many passageways. Araya had no way of recognizing that after 111 years of operation, after millions in gold and copper ore had been wrenched from every corner of the now labyrinthine tunnels, the mine had also been stripped of its support structure. Like a house of cards, the mine was now delicately balanced.

■ ■ ■ ■

Deep inside the San José mine, the miners stripped to the basic necessities — helmet with lamp, water bottle, shorts and MP3 player with a customized dose of Mexican *rancheras,* emotional ballads that chronicle the loves, sacrifice and nobility of the working class. "A lot of times you would see the men working in their boots and their underwear," said Luis Rojas, who worked in the San José mine. "It was just too hot to wear many clothes."

Darío Segovia spent the morning of August 5 attaching metal nets to the roof of the mine — a rustic system to catch falling rocks and prevent men and machines from being crushed. Known as "fortification," Segovia's job was extremely dangerous. He was like a firefighter inside an inferno, attacking small blazes while he was aware the battle was lost. "Before eleven AM, I knew the mine would fall, but they sent us to place the reinforcement nets. We knew the roof was all bad and it would fall. To pass the time we drove the pickup to gather some water at the tanks. It was dangerous; the roof was so fragile."

Mario Sepúlveda missed the bus from Copiapó that morning. At 9 AM he began to hitchhike to the mine. Traffic was sparse and rides impossible along the long road. Sepúlveda was tempted to head back to his cheap boardinghouse. Then a lonely truck arrived on the horizon. When it stopped and picked him up, Sepúlveda felt lucky. He would make it to work after all. At 10 AM he arrived at the mine, checked in, joked with the security guards. By 10:30 AM he was driving into the belly of the mountain.

At 11:30 AM, the mountain cracked. Workers asked the head of mining operations, Carlos Pinilla, what was happening. According to congressional testimony by the miners, Pinilla was heading down into the shaft at the time. He told the miners the sound was a normal "settling of the mine," and kept them deep inside the shaft. Pinilla himself, according to the miners, commandeered the first available vehicle, turned around, and immediately headed for the surface. "He left early that day and he never did that. He would usually leave at one or one-thirty and that day he left around eleven," Jorge Galleguillos testified. "He was scared."

Raúl Bustos knew next to nothing about mining when he entered the San José copper mine on the fateful morning of August 5. Bustos was a man at home on the water, working on boats, repairing, welding and fixing water systems at the Chilean Navy shipyards. He worked there for years until a Sunday morning in February 2010, when he lost not only his job but his entire workplace, which was dragged out to sea by a thirty-foot-high wall of water, a deadly tsunami. The 8.8 earthquake that spawned the tsunami left few factories standing in the coastal city of Talcahuano, so Bustos migrated 800 miles north to the San José mine.

Bustos, forty years old, knew the mine's dangerous reputation but was not worried. His job often kept him in a garage with a zinc-roofed shed, on a treeless hillside repairing vehicles. Sunstroke and homesickness seemed to be his biggest dangers. Every other week, he rode the bus half the length of the nation to see his wife, Carolina. Bustos never complained about the twenty-hour ride or let his wife know his new workplace was so precarious. When a vehicle was reported to have a flat tire and mechanical problems deep inside the mine on the morning of August 5, Bustos stepped inside a pickup and was driven four miles down into the mine, deep into the earth.

The mine was a maze of more than four miles of tunnels. As more than a century of miners had chased the rich veins of gold and copper, the tunnel was not excavated in an orderly fashion, but was a chaotic scene. Loose cables hung from the ceiling. Thick wire mesh was hung from the roof to catch falling rock. Small altars along the narrow main tunnel marked the spots where workers had been killed. In general, the men worked in groups of three or four. Some worked alone. Nearly all of them had ear protection, making it difficult to speak or hear anything but the loud noises of the working mine.

At 1:30 PM on August 5 the miners stopped for lunch, some of them heading down to the refuge where there were benches and they could grab a boost of oxygen. Five minutes sucking oxygen was usually enough to get the men back to work or at least back to the lunch table where they shared a rare communal moment in their solitary world. While they ate, the men fired up "*la talla*" — a distinctly Chilean practice of spontaneous humor that feels like a brilliant combination of stand-up comedy and impromptu rapping. Meanwhile, an entire mountain was sagging above them.

Franklin Lobos was the last man to enter the mine that day — probably the last one ever. As official chauffeur for the mine, Lobos ran an efficient and hilarious shuttle service — entertaining his passengers with wild stories of women and fame as he drove them into the depths of a world that looked like a set from *Lord of the Rings* with its sagging roof, piles of debris, and walls that looked as if they had been carved out by hand a century earlier.

As a former soccer star in Chile, Lobos was a legend. It was like having David Byrne drive you to Heathrow or Mike Tyson as your cabdriver to JFK. Lobos, fifty-three, was now bald, round-faced and low-key. His youthful adventures made him a magical storyteller and he regaled his passengers with the glory days of his career for the soccer club Cobresal. Many of the miners were devout fans, men who grew up watching Lobos score goal after goal as he cemented his reputation on the soccer field.

During his 1981 to 1995 career, Lobos rose to the elite in northern Chile — a demigod who turned free kicks into a one-man show. Even before Lobos touched the ball, the entire stadium was rapt, imagining the impossible trajectory that Lobos would unleash, celebrating his abuse of physics. Lobos's goals were so precise and unbelievable that the Chilean press dubbed him "The Magic Mortar Man," a player capable of half-field bombs that arched exactly to their target. Even Beckham would have applauded. But soccer stars in Chile have an average career of ten years. By his mid-thirties, Lobos was gainfully unemployed and devoid of either the star power or the cash to live up to the legendary status. Lobos tried his luck as a taxi driver, but with two daughters headed to college, he needed cash, and in Copiapó that meant one thing: a job at the San José copper mine.

It was just past 1 PM when Lobos drove Jorge Galleguillos down into the mine inside a cargo truck. Halfway down they stopped to chat with with Raúl "Guatón" ("Fatman") Villegas, who was driving a dump truck filled with rocks and boulders carrying trace amounts of copper and gold. It was then that the mine cracked.

"As we were driving back down, a slab of rock caved in just behind us," Galleguillos later wrote. "It crashed down only a few seconds after we drove past. After that, we were caught in an avalanche of dirt and dust. I couldn't see my hand in front of my face. The tunnel was collapsing." Galleguillos would later compare the scene to the collapse of the World Trade Center. Layer after layer of the tunnel fell in stacks, like pancakes.

As the mine cracked, it unleashed a series of avalanches. Lobos did not dare speed up. Instead he focused on avoiding debris that partially blocked the tunnel. The collapse was now in front and behind him. He crashed into the wall. Unable to see, Galleguillos exited the truck in an effort to guide Lobos down. As the roof continued to rain down, Galleguillos sought refuge in the lee of a water tank. The men finally negotiated a sharp curve and, despite the clouds of dust, slowly began to descend toward the safety refuge.

When Lobos reached his colleagues, they stared at one another in shock. No one could say what had happened. Everyone knew it was unlike any of the mini-avalanches that were habitual inside the San José mine.

One thing they did know: for even the most novice miner, the message was clear — "*El Piston*" was coming. Slumped in the corner of the rescue shelter, huddled behind bumps little bigger than a mattress, all the men braced themselves.

When a mine caves in, the air inside the mine explodes — like a piston — through the tunnels, generating winds so strong that they plaster a working man to the far wall, shatter his bones, crush the breath out of already muddled lungs. "It was like getting boxed in the ears," said Segovia. "It felt like it went through your head."

Small avalanches inside San José were a monthly event, a terrifying but brief rupture that invaded the miners' daily solitude. Even with headphones and the deep bass of reggaeton and Colombian cumbia blasting into their ears, the men never missed the distinctive *craaaaack!* Rock versus rock. Every time it was the same; within seconds each miner sought refuge. The following minutes were guaranteed to bring any one of a possible series of consequences — at best a storm of suffocating dust; at worst, news that a colleague had been crushed. Usually the entire episode lasted a few hours. This was different.

"A true piston effect is like an explosion. It is a deep sound, like a herd of galloping buffalo. You have very little time to react," explained Miguel Fortt, one of Chile's most experienced mine rescue experts. "You can't do much."

"I thought my eyes were going to pop out of my head," said Omar Reygadas, a fifty-six-year-old miner with decades of experience. "My ears exploded." Despite his helmet and ear protection, Reygada was nearly doubled over in pain. Could he even hear? He worried he'd been deafened.

The blast sent Victor Zamora flying. His false teeth, jarred loose, were lost in the rubble. His face was bruised and scratched as waves of compressed air, like miniature sonic booms, battered the men. The air was a tornado-like storm, with rocks and dust firing down the tunnels.

The thick cloud of dust and debris blinded and choked and deafened the group of men, coating them in a layer of dust almost an inch thick, as they struggled to escape the mine, falling, crawling and pushing their way up the mine shaft. Like sailors in a hurricane, they interpreted the energetic blast from Mother Nature as a sign of vengeance from an invisible female goddess — that capricious and omniscient overlord who had last say in their precarious world. Some of the men began to pray.

The force of air shot through the top of the mountain, producing what Araya and others outside the mine described as "a volcano."

Deep in the mine, the men faced a dust storm that flooded their world and was to last for the next six hours. After the roof collapsed, the men were blinded by a cloud of rocks, dirt and traces of the highly precious copper and silver ore that, since the opening of the San José mine in 1889, has lured six generations of miners into this precarious world. "I thought my ears would explode, and we were inside a truck with the windows up," said Franklin Lobos, describing the pressure that damaged the inner ear of his colleague José Ojeda.

Ten minutes after the first collapse, the mountain ruptured again. A short, succinct signal that millions of tons of earth and rock had slipped again. Outside the mine, panic struck.

Mine operators and supervisors who heard the first crack assumed the miners "had burned" — slang for igniting dynamite. Nothing unusual there. But two "burns" in ten minutes? Impossible. The third *craaaaaaaaaaack!* was terrifying and unmistakable. Above and below the mine, hundreds of workers were paralyzed in fear. What was going on down there? Miners never detonated their charges so close together. Curiosity mixed with trepidation pervaded this deserted corner of the Atacama Desert.

Inside the mine, a group of some fifteen miners had battled the dust and struggled to walk up the tunnel in search of safety. They were stopped by a massive rock face blocking the tunnel. The men panicked. "We were huddled like sheep," said José Ojeda. "We heard that sound, I don't know how to describe it . . . It is terrifying, like the rocks are screaming in pain. . . . We tried to advance, but we couldn't; a wall of rock blocked us."

When Florencio Ávalos arrived in a pickup, all the men climbed aboard, stacked like refugees. On the way down they crashed twice, ramming the walls, lost in the dark chaos. As the pickup bounced down the road, one of the miners fell off. Alex Vega reached out and yanked back the flying body, pulling the man to safety. In the chaos he was never sure who he had saved. As he strained to pull the man back to the bed of the truck, something snapped in his lower back. It would be hours before the adrenaline wore off enough for the stabbing pain to begin.

Driving blindly through thick clouds of dirt and debris, it took the men nearly an hour to reach the safety refuge, a shelter carved from the rock. Once they reached the refuge, the men shut the metal doors to block out the dust storm. Then the thirty-three men took turns breathing from oxygen tanks.

The 540-square-foot refuge was little more than a hole in the wall with a ceramic floor, reinforced ceiling, two oxygen tanks, a cabinet filled with long-expired medicine and a tiny stash of food. "The guys would constantly raid the safety shelter so we never knew exactly what was left. They always stole the chocolates and the cookies," said Araya, the paramedic who was also in charge of stocking — and restocking — the safety refuge. "These guys were lucky, though; usually we had only one tank of oxygen in there, but when they got trapped, there were two tanks."

Inside the mine, Luis "Lucho" Urzúa tried to pull the reins tight on his group. Two decades as a miner and a stretch as an amateur soccer coach were enough experience to make leadership a reflex. As shift foreman, Urzúa was the official leader — but the soft-spoken mapmaker had worked less than three months in the mine. He barely knew his troops. Scouring his makeshift refuge, Urzúa took stock of his provisions: ten liters of water, one can of peaches, two cans of peas, one can of salmon, sixteen liters of milk — eight banana-flavored, eight strawberry-flavored — eighteen liters of juice, twenty cans of tuna fish, ninety-six packets of crackers and four cans of beans. Under normal circumstances, the food was meant to satisfy the appetite of ten miners for forty-eight hours. Now there were thirty-three hungry men. "That day many of the guys had left their lunch up at the top of the mine," said miner Mario Sepúlveda. "There was less food than normal."

By four in the afternoon, approximately two and a half hours after the first deep sounds of cracking, the mine was fully collapsed. "It was like a volcano; the hillside spit out debris and from the mouth of the mine there was a cloud of dust," said Araya. He described the sound outside the San José mine when an 800-foot-long section of the mine collapsed: "It wasn't a long sound — more like a final collapse. One deep thump."

The final "thump" that Araya described was a rock estimated at 700,000 tons that sealed the only entrance to the mine. The trapped miners knew the final "thump" was anything but routine, even in a mine as dangerous as San José. The dust alone had nearly killed them, leaving the men coughing, crying and half blinded. Their eyes filled with so much grit that the majority soon developed a crackling hard yellow coating that glued their eyes shut. Even when they opened their eyes, the darkness was impossible to penetrate, and water poured down through the walls.

Instead of their usual battles with dust, the men now faced a muddy, slippery slope outside the refuge. The frequent downpour of stones and boulders echoed like a madman's drum inside a one-mile stretch of rocky caverns where they were now imprisoned. The men lurched awkwardly in the darkness, shutting off their lanterns in an effort to conserve precious battery power.

Their nightmare had begun.

TWO
A DESPERATE SEARCH

Thursday, August 5, 5:40 PM

Mario Segura was wet and cold when he returned to the police station in Copiapó, Chile. After four hours of rescue training in the frigid Pacific Ocean, the wiry commando was ready for a hot shower and a cold beer with his colleague José Ñancucheo. Both Segura and Ñancucheo are members of GOPE, the Chilean *Carabineros* Special Operations Group, an elite police unit trained in everything from dismantling bombs to rappelling down the innards of the hundreds of volcanoes that crown the Andes Mountains — a spine that runs the 2,700-mile length of the Chilean landscape. When an adventure tourist exploring the lip of a volcano in Chile crosses the fine line between extreme adrenaline and sudden slip, these are the guys who search for the remains. When an anarchist bombs a business (a monthly event in Santiago), these are the men sent to the scene.

Highly trained and respected throughout South America as one of the most professional police units on the continent, GOPE members spend many of their waking hours at the gym or shooting range or unraveling disaster scenarios. On August 5, after hours of scuba rescue training, Segura and Ñancucheo were nearing the end of their shift when the phone rang. "I bet it's a rescue," joked Segura, as he sat down for hot tea and sandwiches with his squad mates. As his colleague listened to the call, Segura recognized the instant metamorphosis from relaxed end-of-the-day attitude to mission critical operational mode. The phone call was curt; so were the details. Another mining accident. This one at the San José mine, 27 miles into the hills.

"I looked at my watch when we left. It was six [PM]," said Segura. "I said to Mendez, 'We will be back in three hours.' Rescues always take three hours. I said, '*Compadre,* we'll have our snack when we come back.' I turned off the kettle, but I left the tea ready to be served."

The six men loaded 300-foot coils of rope, gloves, climbing harnesses, and crates of carabiners and helmets with head lamps into their 4x4 Nissan pickup. An orange suitcase packed with a LED lighting kit, similar to those used on professional photo shoots, was also packed in the back, but in the rush the men forgot a key piece of equipment: a tripod stand that allows a rope to be centered over a rescue hole to aid climbers in speedy ascents and descents inside a mine. An error for which one man would later pay dearly.

As the sun dipped lower, the police pickup sped up to the mine — flashing lights parting what passes for evening rush hour in this sparsely populated desert. The men spoke infrequently as they silently ran through rescue procedures in their minds. The drive took only thirty-five minutes but was deceptively dangerous. Sharp curves and an irregular fog that often dumped a slippery and invisible layer of water on the road explains in part why rental cars in the region typically include not only double roll bars and two spare tires but also an extensive first aid kit.

When the GOPE team arrived at the mine, a geologist and geophysicist were waiting to deliver an impromptu lecture in which they sketched the structure of the mine and the estimated position of the trapped miners. Accurate maps were not available on such short notice and guesswork was a major factor in the rescue planning. The geologist was serious and worried. "This is a complicated operation," he told the six GOPE members. "This one will take time." The geologist pointed out a ventilation shaft on the rough map of the mine and suggested the police first find it and, if possible, lower themselves into the bowels of the mountain. There were miles of tunnels to search. Had the men reached the safety shelter near the bottom of the mine? Or the vehicle workshop a quarter-mile higher? With more than a dozen vehicles inside the serpentine tunnels, the commandos prepared for the possibility that the men were alive but trapped inside crushed trucks.

The men, if they were alive, might be anywhere.

At first the administrators at the San José mine were unwilling to acknowledge the scale of the disaster. According to Javier Castillo, a union representative for the mine who said it was his call that first alerted authorities, the management initially banned the men from using company telephones to call for help. Angelica Alvarez, wife of the trapped miner Edison Peña, told a similar tale: "The miners wanted to call down, and since there is no cell phone coverage up at the hill, they asked for a land line. . . . They strictly prohibited them from contacting firefighters, an ambulance or the police. The company wanted to fix this on their own terms." As a growing clan of rock climbers, veteran miners and GOPE commandos studied their options at the mine, family members and ordinary citizens around Chile stared in shock at the newscast. The San José mine had collapsed, and now the names of the miners on duty at the time scrolled down the television screens.

1. Luis Alberto Urzúa Irribarren
2. Florencio Ávalos Silva
3. Renán Anselmo Ávalos Silva
4. Samuel Ávalos Acuña
5. Osmán Isidro Araya Araya
6. Carlos Bugueño Alfaro
7. Pedro Cortez Contreras
8. Carlos Alberto Barrios Contreras
9. Jonny Barrios Rojas
10. Víctor Segovia Rojas
11. Darío Arturo Segovia Rojo
12. Mario Sepúlveda Espinaze
13. Franklin Lobos Ramírez
14. Roberto López Bordones
15. Jorge Galleguillos Orellana
16. Víctor Zamora Bugueño
17. Jimmy Alejandro Sánchez Lagues
18. Omar Orlando Reigada Rojas
19. Ariel Ticona Yáñez
20. Claudio Yáñez Lagos

21. Pablo Rojas Villacorta
22. Juan Carlos Águila Gaeta
23. Juan Andrés Illanes Palma
24. Richard Villarroel Godoy
25. Raúl Enrique Bustos Ibáñez
26. José Henríquez González
27. Edison Peña Villarroel
28. Alex Richard Vega Salazar
29. Daniel Herrera Campos
30. Mario Gómez Heredia
31. Carlos Mamani
32. José Ojeda
33. William Órdenes

For many of the families, the TV broadcast was their first indication that disaster had struck. Not only had the mine owners been slow in alerting relatives, but the list was riddled with errors. Two miners were not on the list — Esteban Rojas and Claudio Acuña. Their families would live hours of distress and shock as they sorted out the truth. So too did the families of William Órdenes and Roberto López, both of whom had been listed as victims but were soon discovered to be safely outside the mine. The informal nature of employment, safety and record keeping at the San José mine was becoming evident by the hour.

Relatives who heard about the collapse

from the national television broadcast began to arrive and clamor for action.

With the magnitude of the rescue mission hitting home, the rescue team was presented with new challenges. How were they supposed to search 2,300 feet deep? Was it possible to evacuate injured men from such a distance? Was the mine even safe enough to enter?

They simultaneously began planning for two options: finding the miners alive or finding them dead. Even in the worst-case scenario, government officials immediately developed contingency plans to evacuate the corpses; a major effort would be made to deliver the bodies to distraught families. Given Chile's notorious trauma between 1973 and 1990, when three thousand citizens were murdered and their bodies "disappeared" under the military dictatorship of Augusto Pinochet, leaving the bodies underground, out of sight and thus "disappeared" from their families, was, simply, never an option.

In the hours since the collapse, attempts by mine workers to enter the mine, first in a truck and then on foot, had proved futile. Headlights and searchlights failed to penetrate the dust-choked air. Thick cracks —

many with water streaming out — illustrated the sheer force of the collapse. The rescuers could see little beyond a massive cloud of dust. The continuing crash of rock slabs slamming to the ground and the eerie groans made by the mountain sounded like a monster being strangled. As the mountain wept, the men dodged the tears. "Miners always say that the mountain is alive, which means that it is moving," said Lieutenant José Luis Villegas, commander of the GOPE unit inside the mine. "They say this because the rocks make a sound like a roar. In this case the whole mountain was roaring."

The entrance to the San José mine is a crudely carved, lopsided rectangular hole, nearly twice as high as it is wide, giving the impression of an open mouth. A rough road slopes gently down into a black abyss, like the passageway into a haunted underworld. Behind the mouth, the body of the mine coils around and around, and for almost four miles, it stretches deep into the earth, like a hidden snake. Seen from a cutaway side view, the mine looks like a boa constrictor — a long, lumpy body with bulges at random intervals.

Beside the entrance stands a battered green sign with the company name — "San Esteban Primera S.A." (or "Saint Stephen the

First") — and a huge drawing of a helmet and work boots with the company's slogan: "Work Dignifies, Doing It Safely Makes It Worthy." Rescue workers passed the sign as they entered the mine and found the floor cracked, the ceiling cracked, the walls split. There were no signs of life but there was abundant evidence of destruction.

In those first chaotic hours when dust still drifted from the mouth of the mine and a winter night's cold air chilled the men, Mario Segura was one of the first policemen to enter the mine. "We went down into the mine and followed it as far as possible, and then we came to a spot where the road was blocked by debris and rock. Usually you find a way around the edges of a cave-in. But this was a smooth rock, like a door that sealed off the shaft," said Segura. "The way the mountain collapsed, even the experts in mining did not understand how so much rock fell. That a whole mountain fell like that? For them it was inexplicable."

The massive chunk of rock that had fallen was not like a dagger, as had originally been thought, but more like a massive ship, roughly 300 feet long, 100 feet thick and 400 feet high. Later estimates would put the weight of the fallen rock at 700,000 tons, nearly twice the weight of the Empire State

Building or, measured in the vocabulary of disasters, 150 times the weight of the *Titanic*. With no possible way to burrow through such a rock, the GOPE commandos explored until they found the ventilation shaft — known as a *chimenea* — and, using rock climbing equipment, they began a slow descent into the still-collapsing, still-groaning mine shaft.

With four policemen keeping an eye on the collapsing roof and anchoring rescue lines, two others lowered themselves slowly into the 6-foot-wide circular shaft. Without the proper tripod stand to guide the flow of rope and prevent it from being frayed on the sharp walls of the shaft, the men improvised. They fixed the ropes to a bumper on their pickup and yanked the lines to keep them from shredding away on the sharp rocks. "There was a *rain* of rocks five meters [sixteen feet] away from us. . . . When it started it sounded like a drizzle of water. Then there was a crash and the whole roof came down. Right next to us," said Segura. "When that *rain* begins, you have to be careful; you are never sure where that rock is going to land."

Under Chilean mining law, every such chimney shaft is required to have an escape ladder. But the San José mine was never a place where safety regulations were strictly observed. One miner, Ivan Toro, remembers that when he started working in the mine in 1985, the standard-issue footwear was sneakers. In September 2001, Toro was sitting down, waiting for a truck to take him topside when a section of the roof collapsed. "We could hear the machines perforating in the level above us when suddenly a slab of rock fell. I was the most affected because it fell on my leg. There were just little strands left and they amputated it. When I arrived at the hospital I lost consciousness," he remembered. The company initially refused to pay because Toro had been sitting on the job. Eventually Toro won his lawsuit, but in Chile's free market economy, the price of a lost leg did little to soothe his trauma. The courts awarded Toro 15 million pesos (adjusted for inflation, roughly $45,000).

The innards of this mountain offered a deadly gamble — the promise of gold versus the risk of death. At first glance, the mine looked definitively dangerous, like a set from an Indiana Jones movie without the snakes. Pools of fetid water. Hidden caves. The roof sagged in spots, and crude mesh nets bolted to the ceiling caught rocks as they fell. The smell inside the mine combined dank humidity with the stink of ammonium nitrate explosives. Clouds of cigarette smoke from the chain-smoking miners were hardly noticeable; in this environment the fantasy of living long enough to die of lung cancer was laughable.

Instead of following standard mining practices and leaving reinforcement pillars throughout the excavated chambers, the San José mine began to resemble a gigantic chunk of Swiss cheese attacked by feral rats. No perfect science exists to explain the collapse of a mountain, but later analysis would suggest that haphazard extraction of valuable copper and gold had stripped the mine of its essential vertebrae. "They even mined away the support pillars," said Vincenot Tobar, former security superintendent at the San José mine. "That can't be. You have to leave support every fifty meters [160 feet]. . . . It is those pillars that prevent a cave-in."

Regardless of the exact mechanics, the policemen were inside a major collapse. And now, without the legally required escape ladders to climb down, the policemen rappelled slowly to the bottom of the 50-foot chimney and found themselves in the main chamber. The rescue workers scanned the otherworldly scene in front of them — casting a cautious eye on the uneven ceiling where rocks hung by what appeared to be invisible threads. The tunnel was 16 feet high and 20 feet wide, large enough for an oversized dump truck to haul out the ore and minerals. With the temperature a constant 90 degrees Fahrenheit and laden with a combined 260 pounds of equipment, the men sweated continuously as they explored the tunnel.

Segura and Villegas were accustomed to gruesome scenes: bomb victims, car wrecks, bloated bodies bobbing at sea. This was another dimension, like a dungeon. The mine was a maze of underground tracks, each leading to deeper mysteries. The vast spaces and curved tunnels created the sensation that life or some living being was just around the corner — out of sight. Nets on the ceiling were filled with rocks, a pitiful attempt to catch loose boulders that now lay scattered across the rugged road.

They were seized by the sensation that they would die, that the mine was a Godzilla-sized monster that could crush them without notice.

"I knew we had to keep searching, but the sound of that mountain, it was like rocks screaming and crying," said Segura, who together with a partner found a second *chimenea* and descended yet another level. At the bottom of the second chimney, as they searched, the two men paused and called out, "Estaaaaaaaaaaaan?!" ("Anyone theeeeere?!") They listened for signs of life. The only response was the slosh of water flowing through newly formed channels and the rattle of collapsing rock. Stagnant deposits of water had ruptured with the collapse and the mine was now alive with a fresh flavor of dirt and debris mixed with 85 percent humidity. In miner lingo, the mine was *asentando,* still settling down.

"Every chimney has a wire mesh to keep the rocks from falling in," explained Segura. "But this was such a huge collapse, debris overflowed the chimney. We were going down the final chimney and near the bottom it was filled with rocks."

"We all were very anxious. We reached a lower level and we were a little upset that the tunnel was still blocked but that kept us going," said Lieutenant Villegas, the GOPE commander. "We said, 'No . . . the next one will be open.' . . . And we kept on going down, but each level was the same [sealed off]." As a second team of police commandos searched for a way to bypass the sealed ventilation shaft, the mine unleashed a shower of rocks. The effect was like a spit in the face. "While they were climbing down another landslide hit, blocking the ventilation shaft," said Villegas. "After that, the entrance through the ventilation shaft was impossible."

While the police commandos searched and explored the mine looking for the trapped men, word of the accident spread — a cave-in at San José, thirty-three men inside. Rumors flourished, including the version that twenty-two miners were crushed and dead. "I heard they found my father's truck, with blood inside it, and he was dead. I cried and cried all that day," said Carolina Lobos, twenty-five, daughter of Franklin Lobos, the former soccer star trapped in the mine. "I had no more tears left, yet still I cried."

Dr. Jorge Diaz, medical director for the Asociación Chilena de Seguridad (ACHS), was on call at a Copiapó medical clinic. Upon hearing of the San José mine collapse, he immediately cleared out the hospital beds, summoned his staff and prepared to treat the injured. No one ever arrived.

After her unsuccessful attempt to convince her husband, Mario Gómez, to stay in bed and not go to work that fateful morning, Lillian Ramírez waited nervously at home. At 8 PM, when she heard the sound of the truck that delivered her husband home, she put the dinner in the microwave. "It was very odd that my husband was taking so long to come in. So I opened the curtains and saw my husband's boss. . . . It was very strange. I put my hands to my face and said, 'My God, something has happened.'" The mining executive asked Ramírez to come with him and said there had been a small accident that would be resolved the following day. He refused to provide details, provoking further panic for Lillian Ramírez, who immediately sought out her nephew and drove up to the mine. It would be months before she returned home.

Surrounding the entrance to the San José mine, the golden brown desert was buried by acres of gray debris, sharp piles of rocks known to miners as "sterile material." As they held no signs of the rich veins of gold or copper, these rocks were dumped in the desert, decades of detritus tossed in uneven waves across the landscape.

These "sterile" rocks were now windbreaks and refuge for the growing clan of families streaming up the hill. Tiny altars with a single photo and candles were now accompanied by signs such as *"Fuerza Mineros — Los Estaban Esperando"* ("Be Strong, Miners — We Are Waiting for You"). The individual face of Jimmy Sánchez, a grim photo taken from his job application, stared out mutely, a silent scream. On an adjacent rock an orange miner's helmet was propped up, sheltering two lit candles beneath it.

Dozens, then hundreds, of family members and friends of the trapped miners flocked to the San José mine on the evening of August 5 and the dawn of Friday, August 6. They brought sleeping bags, food and cigarettes, which they consumed incessantly as they gathered nervously near the mouth of the mine. "I know he can survive this, he once lived as a stowaway on a cargo ship and went twelve days without eating," said Rossana Gómez, twenty-eight, as she proudly described her father, Mario Gómez, the oldest of the trapped men. "He survived *el accidente,*" she said in a coded reference to a dynamite mistake that shredded her father's left hand years ago, splaying it throughout a mining camp. "I am sending him tranquility and comfort," she added, confident that her father would be saved and comforted by her conviction that Dad's long-time desire to mentor a young son would be fulfilled.

Gómez — his lungs failing, his seven fingers proof that he was a veteran — was a survivor who could adapt to this dark dungeon of a life and might adopt the younger, weaker miners. Gómez could teach the young pups the art of survival on the job.

Rossana proudly promoted her father, saying, "He provides strength to his mates."

If anyone needed reinforcement, it was the Bolivian-born Carlos Mamani — the only non-Chilean in the group. August 5 was Mamani's first day of work, a single shift he had picked up as a moonlighter, an extra job to help defray the cost of caring for eleven-month-old Emili, his new baby. Now Mamani was trapped. Given the century-long animosity between Chile and Bolivia, living in a hole some 2,300 feet deep, surrounded by thirty-two Chileans was, for a Bolivian like Mamani, akin to being a Serb stuck in a Croat foxhole.

Of the thirty-three men, twenty-four lived in the nearest city, Copiapó, a mining town with a population of one hundred and twenty-five thousand, where an estimated 70 percent of the local economy relied on the mines. News of a mining accident surprised few of the locals, many of whom were third-generation miners. The local Copiapó newspaper, *El Atacameño,* regularly ran headlines about crushed and dismembered miners. Yet instantly this felt different. The depth at which the collapse had happened and the number of victims were noteworthy, even in a community fluent in the language of mining tragedies.

■ ■ ■ ■

The desert mountains and salt flats of northern Chile are loaded with such riches that slightly over half of the Chilean export revenue comes from mining. In a good month, the nation exports nearly $4 billion in copper. Fully one-third of the world's copper comes from Chile, and the tale of the "green gold" has figured in the nation's economic success for the past two decades, a boom that has not included many of the communities in the mining area. With copper prices tripling from roughly $1.20 a pound to well over $3.00 over the past five years, old tailings and second-rate mines were reevaluated. What was junk at $1.20 a pound might well be profitable if copper prices remained above the $2.50 price level. Abandoned mines and older, more dangerous operations were suddenly valuable and viable.

The Atacama region is home to major mining operations but it also has the second highest unemployment rate in Chile. While copper companies in Chile earned an estimated $20 billion in profits in 2009, government statistics showed the area to have among the fastest rise in poverty in all of Chile. "In other words, one of the richest regions in the country is, at the same time, one of the poorest," concluded an article in *The Clinic,* an alternative newsweekly based in Santiago.

As they gathered at the mine, the disparate families shared a common anger — the accident had been so widely predicted, it was overdue. Yessica Chilla, partner to Darío Segovia, forty-eight, one of the trapped men, remembered, "The day before the accident, he told me the mine was about to settle and that he didn't want to be on the shift when the collapse arrived. But we needed the money. His shift had ended, but then they offered him extra hours. No one refuses because they pay you double. That day he was going to earn ninety thousand pesos [$175]. But he wanted to leave this job to run a trucking business."

Elvira Katty Valdivia didn't hear about the accident until many hours after the collapse: "A friend from college called me. 'Katty, do you know about what happened? It seems Mario is on the list of those trapped inside the mine.' She told me to turn on the TV and I started watching and I saw the list. There was Mario Sepúlveda." Valdivia's dark skin, straight black hair and penetrating gaze highlighted a beauty that had been sorely tested in recent weeks. With her laptop propped up inside a nearby tent, she attempted to maintain the clients for her accounting business. While she balanced the books, her life could not have been more off-kilter. A hole drilled from her feet, with precise luck, would have pierced the tunnel where her husband, Mario, fought, battled and prayed in his struggle to survive. "I feel very sorry for him. Me here, and him down there, seven hundred meters [2,300 feet] deep," she said, indicating the ground. "I would like to be with him, be able to touch him, tell him that I love him very much." Valdivia expressed bitterness toward the mine owners. "They never told me anything. They didn't tell anyone. They didn't tell us a member of our family was down there trapped in the mine."

Valdivia's employer — U.S. accounting firm Price Waterhouse — assured her her salary would be paid in full while she maintained a vigil on this remote outpost as she awaited news on the fate of her husband. With two teenage children, Scarlette, eighteen, and Francisco, thirteen, in tow, Valdivia began to organize her life from inside her temporary home at the mine site. With rumors of death and entrapment swirling in her head, Valdivia watched as her world disintegrated. "People were running everywhere and screaming," she said. "My son was crying and I was trying to console him. It was a very difficult moment. . . . I couldn't sleep. I asked myself, Why me? Why me? Why is this happening to us?"

Chilean President Sebastian Piñera was in Quito, Ecuador, when he first heard of the mining tragedy. He could be excused for thinking the same as Katty Valdivia: "Why me? Why is this happening to us?" It was Piñera's second successive tragedy in his short tenure as president. When he took office just four months earlier, Piñera inherited a nation shattered by the February 27, 2010, earthquake. That quake left hundreds of thousands homeless and hundreds more dead when a tsunami crushed the coastline. Piñera's ambitious political agenda was also leveled by the 8.8 Richter scale quake, the fifth largest ever recorded. Instead of a fresh slate to highlight new ideas, Piñera's team was dealing with thousands of collapsed adobe structures, destroyed hospitals and the wreckage along an estimated 1,200 miles of Chile's modern highways.

"I was with President Correa in Ecuador," said Piñera. "Our diagnosis that first evening was clear. We knew there were thirty-three men. They were trapped at seven hundred meters [almost half a mile] and after a diagnosis of the company, it was seen as a precarious situation. There was no possibility for them to respond. The option was thus very simple. The government would assume responsibility for the rescue or nobody would. It was much more simple than people think."

Piñera ditched protocol, canceled a strategically important reunion with Juan Manuel Santos, the newly elected Colombian president, and rushed back to Chile. He ordered top aides to the scene that very night.

In an effort both caring and self-serving, the Piñera administration saw the crisis as a perfect stage to highlight the can-do attitude of the nation's first elected right-wing government in half a century. Piñera bet his dwindling political capital on the fate of thirty-three unknown miners. It was a gamble that would later reinforce the billionaire businessman's reputation as a brilliant short-term stock trader.

Day 2: Saturday, August 7

The men had now been trapped for two full days, yet no sign of life had been found. Basic, primal fears began to haunt the rescuers. Did the men have air? Were they injured and dying slowly? How would they eat?

Below ground the rescue effort hit another

setback. The rescue workers had been trying to find a route around the blocked ventilation shafts, but with the mine still shifting, the shafts began to collapse. The massive battleship-sized rock slipped a fraction, sending more small avalanches through the mine. Now the GOPE mission changed from rescuing trapped miners to evacuating the rescuers and avoiding a second entrapment. Without the tripod to guide the rope, the police rushed to extricate their colleagues who were being bombarded by rocks. If they pulled too fast or to one side, they risked slicing the rope and delivering a rescuer to his death. If they went too slow, the chance of a large rock knocking him unconscious grew by the second.

"We train for this. We have to study geology, and part of our curriculum is mine rescues," said Hernan Puga, a GOPE member who mentioned that the local mountains housed an estimated two thousand small-scale mines. He compared the vertical descent and ascent to the type of training the police regularly carried out for special operations inside prisons.

When all the rescue workers had been pulled free, instead of celebrating their near escape from death, the police commandos were filled with frustration.

"They were very upset," said Commander Villegas. "We were frustrated, but that changed when we had contact with the families. The hope and faith they had encouraged us."

Chilean mining minister Laurence Golborne arrived at the mine on Saturday. He had struggled to find commercial flights back to Chile, and so was picked up in Lima, Peru, by the Chilean Air Force and flown to the mine. Upon arrival, Golborne was stunned by the disarray he found. Clearly, the mining executives running the San José operation were overwhelmed and undercapitalized for a major rescue effort. After taking the lay of the land, Golborne was proud to inform Piñera that he had organized the arrival of the first drilling rig. The president was unimpressed. "Okay, well done. Now I want you to get not just one but ten drilling rigs," he told Golborne. The president's obsession with maintaining multiple rescue options would become a hallmark of Operation San Lorenzo.

The rescuers told Golborne they had hope the men might be alive. Despite the rumors, no evidence of crushed vehicles or broken bodies could support the fear that the men had been wiped out in a single crushing blow. Daily routine inside the mine was predictable enough to deduce that when the collapse hit, the men were in the lower reaches of the mine and at least some of the group could still be alive in the blocked tunnels.

"We knew the miners had enough water because during drilling they need to have big tanks of water. The problem was the oxygen," said Golborne. "When the shaft collapsed, we really felt angry and powerless. We informed the relatives about this collapse and that we couldn't carry out a traditional rescue [via the mouth of the mine]. . . . I didn't try to give them false hopes. I committed myself to tell them only the truth. I didn't want to cause any gossiping. In this type of situation people talk a lot. You could expect people to say they were all dead."

Golborne's announcement to the families was brutally honest. He told them the rescue effort was suspended. He broke down and cried in front of the families as he announced, "The news is not good." Then rescue workers packed up and began leaving. Firefighters, rock climbers and the GOPE police began to exit the mountaintop. Segura and Ñancucheo were discouraged and humbled. They had been certain they could rescue the men.

"When I saw the GOPE guys go, the rescuers go, I thought if they are going, it is because the miners are all dead," said Carolina Lobos. "I cried. We all cried."

"I felt helpless and desperate," said Lillian Ramírez. "All the relatives went on strike and wanted to get the mine bosses with wooden sticks — like vandals. We made a human chain and told them that we were not going to let anyone leave the mine. The anger and desperation made me push a policeman. . . . Then I realized it was a mistake, what we did, but desperation makes you do many things. And to recognize that is human. We really did not know what was happening."

Pablo Ramírez protested. A shift supervisor at the San José mine who had been among the first to volunteer for the dangerous rescue operations, he insisted they needed to push forward, to continue looking for the men. Ramírez was sure that on one of his missions deep inside the mine he had heard the bleat of truck horns. His rescue colleagues ridiculed him. "No one believed me," he said. "They said it was the souls of the dead miners haunting me."

Three
Stuck in Hell

Thursday, August 5 — Afternoon

Pablo Rojas had arrived at the San José mine that morning with such a hangover that as soon as his group had finished reinforcing walls and propping up the ceiling, he went to lie down in the peace and quiet of the safety shelter, 2,250 feet deep, near the bottom of the mine. Rojas's father had died days earlier and even before the big night out, his head had ached with pain. The massive cave-in roused the bedraggled Rojas, but he was slow to appreciate the magnitude of the disaster.

Claudio Yañez had been preparing to set dynamite charges when the blasts of air from the collapse nearly knocked him over. Yañez was among the first group to arrive at the shelter, and he watched the other miners struggling to do the same, as the mine continued to heave and shift. "They arrived little by little," he said. "The guys came down to try and use the telephone, but it didn't work. We looked at each other for the first time and in desperation. We couldn't believe what was happening."

Raúl Bustos had been working inside a mechanic's workshop just up the tunnel from the shelter when the collapse hit. In a letter he later wrote to his wife, he described the scene: "The suction and the air knocked us all over."

Inside the shelter, best friends and relatives sought to find out who had survived. Florencio Ávalos, thirty-one years old, found his twenty-seven-year-old brother, Renán. Florencio felt paternally responsible; he had encouraged his younger brother to work at San José. Neither of them saw it as a career, but compared with the alternative of seasonal work picking grapes in their tiny pueblo in the mountains near the Argentine border, the job here was quite literally a gold mine.

Esteban Rojas hugged his three cousins — a gracious thanks that they were all alive. Best friends Pedro Cortés and Carlos Bugueño also celebrated their survival; neighbors since childhood, they were inseparable and had started work at the mine on the same day.

Franklin Lobos, however, was distraught. As the driver of the last vehicle into the mine, Lobos had passed Raúl Villegas's truck rumbling up the ramp. Calculating the time of the collapse and the estimated position of the truck, Lobos feared the worst and could practically see the crushed vehicle in his mind's eye. Given the massive collapse of rock, the men had little doubt that *la mina maldita* (the cursed mine) had stolen the life of another colleague.

Lobos knew the shelter well; one of his many tasks was restocking the safety shelter. He had never liked working in the mine. At his previous mining job he had been trapped by a cloud of smoke and forced to retreat to the bottom of the mine to avoid suffocating. For eight hours, as his family gathered outside, Lobos and his colleagues had wondered if they would be given a second chance to live. Now Lobos was looking for a third chance.

All thirty-three men had somehow survived the massive collapse. Several were bruised and a few bloodied but not one had a broken bone. No one was missing.

Inside the shelter, Luis Urzúa, the highest-ranking man in the mine, sought to control the men. As shift foreman, Urzúa was not required to participate in the physical work but instead guided, prodded and motivated the men under his command. In the hierarchical world of Chilean mining, the foreman is absolute leader, his word followed with military discipline. Questioning an order from the shift foreman was sufficient reason to be disciplined or dismissed. "The world of natural selection functions quite strongly in this environment," explained Dr. Jaime Mañalich, the Chilean minister of health. "To arrive at the position of shift foreman, you have to pass through many a test."

Urzúa was a solidly built man with soft eyes and a leadership style built not on being a brute but on most often being right. With more than two decades' experience inside mines, Lucho Urzúa had the experience to command his troops, but he was a recent arrival to San José. That he had worked there for less than three months now hung heavily in the tense and dirty air inside the safety refuge. The men questioned his ability to coordinate the disaster response. Why should he be the leader? Did he even know the mine? Urzúa did little to garner support when he suggested the men stay in the refuge, confident that a rescue operation would save them. In the first few hours after the collapse, raging arguments erupted. Tempers flared. Urzúa was losing control.

■ ■ ■ ■

Imprisoned in the shelter, Sepúlveda was calm as he paced about. He had practically predicted this very collapse. How many times had he argued with labor and safety inspectors back in Copiapó? He had spent days encouraging, haranguing and berating them to investigate the San José mine for safety violations. Sepúlveda had attempted to form his buddies into a workers' union, but gave up in frustration when he came to believe that the representatives of the Central Unitaria de Trabajadores (CUT), the national workers' union, were as self-serving as the mine owners. According to Sepúlveda and other miners, the union was in the pocket of the mine owners, spending more time breaking the union than fortifying it.

Sepúlveda, a short, balding man with a wide crooked-tooth grin, was a workaholic who combined a love of physical labor and an unbreakable spirit. To his colleagues he was either El Perry (Chilean slang for "the Good Dude") or El Loco ("The Crazy One"), the unofficial mine shaft jester. He would regularly launch sharp jibes at the mine management, but always with such spontaneous humor that even the targets of his barbs would find themselves laughing along. At the end of a typical day's shift, as the miners took the twenty-five-minute journey in a truck that spiraled up 10 miles per hour from the bottom of the mine, Sepúlveda always had a captive audience: his exhausted colleagues. They marveled and cheered as he improvised monologues and skits. Who else but El Perry would pole dance on the bus as the miners left work? A natural mimic and charismatic character, Sepúlveda was in a situation his hyperactive nature found oppressive — enclosure. He was desperate to find a way out.

Sepúlveda and Mario Gómez organized the miners into three separate missions. Even as the mountain roared and the dust billowed around them, the men began to scour the mine for escape routes. Food, air and clean water were all limited, and the mine continued to rumble and send signals of another monstrous collapse. It was clear they would all die without swift action.

The main shaft of the mine was a ragged tunnel with uneven walls that, when lit by vehicle headlights, sent shadows bouncing about. It looked like the bowels of a haunted world. Side tunnels, caverns and storage rooms had been carved at seemingly random spots. Huge tanks of water were stashed throughout the mountain. Containing as much as 4,000 gallons each, the water was used to operate drilling machinery inside the mine. Had the men been able to see the mine from a cutaway side view, it would have resembled an anthill riddled with shafts.

The levels of the mine were measured in meters above sea level. Given that the entrance to the mine was roughly 800 meters (2,600 feet) above the ocean, the very bottom of the mine was called level 45. The refuge shelter where the men were gathered was level 90. The thirty-three men were trapped near the very bottom of a vast mine.

Secure in their faith that rescue teams were already mobilized, the men were desperate to send a message that they were still alive. Some of the miners began gathering truck tires and dirty oil filters. Richard Villarroel, a twenty-seven-year-old mechanic working as a subcontractor in the mine, was sent in a pickup truck to drive up the tunnel. He arrived at level 350 where the tunnel was sealed shut by the block of rock. Villarroel looked for cracks in the rock, then stuffed the holes with rubber tires and oil filters, which he ignited. Thick black clouds of smoke filled the tunnel, enough of it seeping upward, he hoped, to alert the rescue teams to their location.

A second group of miners gathered sticks of dynamite to detonate the charges in a brief yet distinctive explosion that would, it was hoped, be heard by rescue workers. Other men began to scour the new configuration of the mine to find pockets of air.

Urzúa, a trained topographer, began to sketch a map, a crude attempt to take the dimensions of his new reality. Commandeering a white pickup as his office, Urzúa began his mapmaking in earnest.

While some men still respected his leadership, there were notable exceptions. Juan Illanes, a fifty-two-year-old subcontractor, emboldened by his experience as a soldier in Patagonia, where he had spent nearly two years in a foxhole, considered himself exempt from Urzúa's chain of command. Illanes and four other workers hired to maintain and operate vehicles inside the mine were not mine employees. This meant that in the norms of a Chilean mine, Illanes and his group were second-class citizens. A tribe apart.

Without light, there was no day. Or night. Every routine was destroyed, eliminated or radically altered. As their head lamps began to run out of battery power, the men used them sparingly. They entered the fragile world of sensory deprivation. Add in the emotional overload from a near-death experience and it makes sense that the miners lost all notion of time. The veteran miners understood immediately the technical challenges of drilling and hacking through hundreds of feet of solid rock. For them, the rescue — if it ever came — was a complicated and uncertain operation.

Psychologists understand that in such circumstances, the individual survival instinct trumps the common good. Adrenaline pumps into the brain and survival chemicals flood the body, enabling remarkable feats of physical strength but also a single-mindedness that blinds the miners to the value of stopping for a moment and making a plan. As those first hours passed, the thirty-three miners began to act like a roaming band of hungry animals, haphazardly shitting and urinating throughout their reduced world. Ignoring calls for group unity, they set up disparate caves in random corners of the tunnel. Few of the men slept that first night.

Day 1: Friday, August 6

Having huddled through the night on cardboard strips, in an attempt to stay dry and to blunt the sharp rocks, the miners arose wet and anxious. José Henríquez sought to begin the new day with a dose of hope: a collective prayer. The round-faced, cheery fifty-four-year-old worked in the mine as a *jumbero,* an operator of heavy machinery, which was among the highest-paid jobs in the mine. But that was his day job. Henríquez's passion was preaching the miraculous powers of Jesus Christ to his congregation in the southern Chilean city of Talca. Gathering the men in the refuge, Henríquez gave a brief prayer — enough, it seemed, to relax the men and allow Lucho Urzúa and Mario Sepúlveda to organize a mission. Claudio Yañez had a Casio wristwatch, allowing the men to reorient their schedule and day. "I didn't need a watch down there," said Sepúlveda. "You know what works as a clock? My stomach. I could tell what time it was by what I wanted to eat. Your body does not react the same to the idea of a steak at seven in the morning as it does at seven at night."

Many of the miners were convinced that they should remain in the shelter and await a rescue. Sepúlveda summed up his thoughts on that strategy in a very public, very loud and succinct opinion: that's suicide. Sepúlveda wanted, needed and demanded action. His entire character was a whirl of energy and proactive survival. From childhood on, his life had been a fight to survive.

His mother had died giving birth to him and he had been abandoned by his father. Young Mario grew up sharing a bed with six other siblings. At times he slept in the barn alongside the livestock, even eating the animals' food to survive. "I was very, very poor and they treated me worse than the animals," said Sepúlveda. For the now-middle-class thirty-nine-year-old with a wife and two teenage children, escape from the mine was the very mission for which he felt his life had been preparing him.

The miners divided up into separate groups. One team used heavy machinery to create noise. Despite the massive collapse, the men had at their disposal a flotilla of vehicles ranging from pickups to the Jumbo, a 30-foot-long truck with a drilling platform on the front end used to perforate the roof and make holes for dynamite. The men moved all the vehicles to the highest point of the tunnel. Once astride the blockage, they began to create a cacophony of sounds. Honking horns. Exploding dynamite. Bashing huge metal plates against the bulldozer. The short crack of dynamite and the echoing metallic clang reverberated through the tunnel, but was it enough to be heard? Would at least one member of the rescue team be alerted? The men continued to attack the roof of the mine with the Jumbo — like a mad woodpecker, the machine pecked wildly, making an infernal racket.

"We used the trucks to smash against the walls," said Samuel Ávalos. "We connected the horns on the truck to tubes that ran up to the surface so we would be heard above. We took turns screaming into those tubes. . . . We were desperate."

Alex Vega wanted to climb out of the mountain by following a series of cracks that led, he guessed, all the way to the surface. He was convinced that an escape path was possible but the men had limited battery power on their lamps and no way to carry enough water for what might be a day-long expedition. "We were afraid of getting crushed by falling rock," he said. "There was a chance of being trapped."

A second team of miners, led by Sepúlveda and Raúl Bustos, scouted an escape route via a ventilation duct. This chimney — one of an estimated dozen air ducts that made the air in the mine nearly breathable — rose vertically for 80 feet. "We started to look for alternatives; we climbed up 30 meters [100 feet] on a hanging ladder. We reached the level 210 and saw that it was also blocked," Bustos wrote his wife in a letter later. "There was another chimney but it did not have a ladder."

In many Chilean mines, every chimney would have been a clean circle, shooting up like a skylight to the next level of the mine and lined with safety equipment ranging from a ladder to escape lights. Apart from providing a vent for air to circulate inside the mine, the chimneys are designed to provide an adequate secondary escape route if a tunnel collapses. In the San José mine, the second chimney shaft was unlit and the ladder decrepit. Furthermore, the chimney was astride the main tunnel, meaning that a single accident could simultaneously wipe out both escape routes. It was a basic failure that the miner's union, led by Javier Castillo, had denounced for years. The trapped miners now understood his logic.

Sepúlveda scouted the chimney and decided an ascent was risky but possible. A cascade of rocks was ricocheting down the tube — but he had a helmet. He adjusted the lamp on his helmet skyward and began slowly advancing. The ladder was designed for just such an escape effort but decades of constant humidity had eaten away the rungs. As he reached up, Sepúlveda could feel them giving way. Some metal rungs were missing. Like a desperate rock climber, Sepúlveda began to improvise. The tunnel was four feet wide, far too big for him to brace a leg on each side. So, grabbing a plastic tube that ran the length of the chimney, he tried to find a nub and a foothold on the slippery stones. Meanwhile, a constant hail of rubble continued to clang down on his head. The mountain was still crying, peeling apart. Determined to claw his way out, Sepúlveda summoned his muscles to obey. He reached his hand up and had begun to pull his body up when he slipped. A rock crashed into his face, slicing his lip and knocking out a tooth. Another rock, this one the size of a tennis ball, whooshed by. Sepúlveda had cheated death by a few inches. When yet another rock bounced harmlessly by, Sepúlveda took this as both an omen and a cue to retreat.

"I felt like a twelve-year-old, so strong, so much energy. I never got tired. The only thing I wanted was to get out," said Sepúlveda, who described his experience in mystical terms. "In the middle of this chimney, I felt this was divine . . . my hairs stood on end. Something told me 'I am with you.'"

Sepúlveda felt an overwhelming joy and confidence as he descended the chimney. "I came back and told them no one will die here, those who want to believe, it is up to you, but if you believe, hold God's hand and mine and we will get out of here."

The reaction to life-altering trauma evolved in idiosyncratic ways for each individual — depending on each miner's personality. In drastic experiences like the San José mine collapse — which psychologists define as "situations of extreme confinement" — some victims wilt. Others bloom. For Mario Sepúlveda, his entire life seemed to be geared toward just such a challenge.

He relished his new emerging role: leader of the pack.

Day 2: Saturday, August 7

With no communication from any rescue team, the miners spent another restless and fearful night. In the morning, the men agreed to pray again with Henríquez. A semblance of routine had begun to form by at least gathering to pray together, but desperation was beginning to take hold. Food was running short. The 10 liters (10 1/2 quarts) of bottled water were not nearly enough, and the men began to drink from the huge 5,000-liter (1,300-gallon) tanks usually reserved for industrial drilling machines. The water in the tanks was months old, filled with dirt and grime. "We drank it but it tasted like oil," said Richard Villarroel.

Claudio Yañez drank and drank the dirty water — up to 7 liters (almost 2 gallons) a day. The taste reminded him of diesel fuel and dust. He knew the water was filled with mineral residue and had been stagnant for nearly half a year, but the thirst was brutal. And so Yañez continued to drink.

"The hierarchy was lost almost immediately," said Alex Vega, who worked as a mechanic and knew the mine intimately after nearly a decade inside. "The thirty-three of us were one and we began a democratic system; the best idea that made the most sense was the idea that ruled."

The men began to vote on nearly every important decision. At noon they held a group meeting that combined the democratic debate of a New England town meeting with the humor of the British parliament. Ideas were put forward and either immediately ridiculed to death or debated openly. All the men had an equal voice. Ideas were measured by their intrinsic value, regardless of whether it was sponsored by the shift foreman or the lowliest assistant.

The miners had now spent nearly two full days underground. The batteries on the men's lanterns were fading. Cell phones were now dead. Though there had never been cell phone coverage in the shelter, the men used them as lights, clocks and speakers, listening to music to soothe the pain of the deep silence.

Some of the younger, less experienced miners began to panic. Nineteen-year-old Jimmy Sánchez, the youngest of all the miners, began to hallucinate. He imagined his mother coming to visit him deep in the mine, and in his dreams she brought fresh *empanadas* — a Chilean meat pie flavored with onion and adorned with a single black olive. As a quick lunch snack, the *empanada* is like most of the food in Chile — forgettable. But for Jimmy and his mates, at this underground altitude, even the fresh memory of an *empanada* was food for the gods.

Other miners, unable to deal with the emotional impact of their ordeal, simply froze. "They stayed on their bed all day; they never got up," said Villarroel. Time passed excruciatingly slowly for the men, a massive silence filling the gap. No drilling. No sounds of dynamite. Not a single sound from above. Just the torturous drumbeat of water and falling rocks.

The men repeatedly walked hundreds of yards up the sloping, curved tunnel to stare in depressed shock at the massive boulders. Though they were sure that rescuers must be searching above, the sound of silence was terrifying, and a faint thought began to grow. Will we ever get out of here?

They would curse the rock, "*Piedra maldita, concha de su madre!*" ("Damn rock, your mama's pussy!") The other miners would rally their enthusiasm for a brief cheer of "*Viva Chile*" ("Long live Chile") but then trudge back to the refuge with the same message — no news.

The men needed a miracle — and food. After just two days, their bodies were beginning to shrivel, and their faces became gaunt as their energy began to ebb. Shadows of whiskers began to shroud their faces and dirty hair poked out in stiff clumps. As they spoke face to face, the disintegration of civility was evident. Smells of sweat and humid humanity became so intense, the men began to abandon the shelter and sleep on the rocky tunnel floor.

The men began to break up into groups. Fighting broke out over cardboard. Subclans formed as relatives and old acquaintances bonded in a sense of survival. The leaders, including Sepúlveda and Urzúa, settled at a bend in the tunnel 105 meters above sea level [350 feet]. It was instantly baptized "The 105 Group" or simply "105." These men had the best air, a floor that was less wet, and breathing room from the other two groups. Farther below, another group moved into the Safety Refuge and called themselves *Refugio*. Its hard ceramic floor made sleeping difficult inside, but the roof was reinforced with bolts and a metal mesh to catch falling rocks.

A third group was essentially left to fend for themselves. Cousins Esteban Rojas and Pablo Rojas, plus Ariel Ticona, who had married into the family, formed a clique here at the most dangerous sleeping spot. Just outside the refuge on the mine's main road a second site became known as "the ramp" or *Rampa*. This sleeping area was less claustrophobic, as air blew lightly through the tunnel. But the drawbacks were notable — the area was wet, meaning the men barely slept and at times had to build canoe-like shelters to keep the flowing water at bay.

The cardboard did little to stave off the dampness, the moisture was incessant and they could neither sleep nor stay dry. Some of the men began to sleep in the beds of the trucks. "We did not have much hope for a speedy rescue," said Alex Vega, "and the hardest wait began, in silence and with no certainty of what would happen to us."

Day 3: Sunday, August 8

By 6:30 AM on Day 3 the men were awake and ready for prayer. Henríquez was cheerful and promised that God would respond to their prayers. Every day that passed, his sermons and prayers felt like a lifeline, a single feature to grab on to and hold tight. The rescue might or might not be coming close, but the miner's faith was helping sustain them. They began to refer to Jesus as "the thirty-fourth miner."

After prayers, Mario Sepúlveda roused them for a group meeting. Sepúlveda's demeanor injected the men with his trademark enthusiasm without betraying the miner hierarchy. He lectured the group to respect Urzúa. If the leader did not want to lead, then Sepúlveda would gladly take the lead as a man capable of cajoling, threatening and motivating the men as a positive force.

Despite flagging energy, individual skills began to shine. Raúl Bustos, the survivor of the huge earthquake, drafted Carlos Mamani, the young Bolivian, to help him build a series of canals to help drain off the water running through their camp. Edison Peña rigged up a system of lighting — using the batteries from the vehicles, in particular the bulldozer-like truck known as the "scoop," which had a 220-volt outlet built into the chassis. Instead of relying on intermittent and dim lanterns, Peña's work brought the men a constant beam of light. Illanes also designed a system to charge the head lamps by connecting them to the vehicle batteries. For hot tea, the men boiled water by running one of the trucks and sticking half-liter bottles of water around the exhaust pipe. Though the plastic was too hot to touch, it never melted and the warm water mixed with a few bags of tea the men found provided a small moment of comfort. The men also placed wet boots and wet clothes over the motor and used the heat from the engine as a clothes dryer.

Improvised baths were taken in a nearby mud pit. Soap was nonexistent, as were the basics of hygiene, including shampoo and toothpaste. For toilets, the men used an empty oil drum. When it was near full, they shoveled dirt and gravel atop the stinking mess and dumped it in an area downstream from their camp — covering the waste with more gravel. Still the smells began to rise up in nauseating waves. Victor Zamora could not stomach the stench. He moved out of the shelter to sleep on *la rampa*. For Zamora the entrapment was "a nightmare . . . we did not know if we would ever make it out." To escape the daily horror, Zamora began to keep a diary, a written record of his experience. His literary interests surged as he started writing poetry, short stanzas of optimism and survival cut short only when he ran out of ink.

Food from the shelter was now under strict guard. Only Luis Urzúa and Mario Sepúlveda had access. They proposed a strict rationing program, a plan that was quickly agreed upon after a simple democratic exercise: the men voted. "Sixteen plus one was majority," explained Urzúa. "We voted on everything." The men agreed to eat once a day, little more than a sliver of food. "We would take a spoonful of tuna fish, maybe half the size of a bottle cap, and that was our food," said Richard Villarroel. "Our bodies were starting to be consumed."

What little food had been left in the shelter was fast disappearing. Half the cartons of milk were long expired. The heat had curdled the contents into banana-flavored clots. It was rancid.

Claudio Acuña sniffed the carton. "Smells okay," he thought, then without hesitation he chewed and swallowed a full liter of the chunky milk.

Samuel Ávalos scoured the mountain for scraps of food. "I turned over the garbage bins, searched through it all but it was filled with papers, reports from the mine." In the bottom of each of six bottles of Coca-Cola, he found there was a flavorful drop of drink. He found orange peels and ate them gladly.

Mario Gómez, the veteran miner and former merchant marine, encouraged the men to hold out. He described a journey he had made as a young man when he had hidden aboard a Brazilian cargo ship. For eleven days, young Gómez had lived inside a life raft, a stowaway who survived on rainwater and little else. "We will survive," Gómez would remind the men. Gómez carried undisputed seniority status. He first worked in the San José mine in 1964, before some of these colleagues were born. He had seen the mine grow from a small pickax-and-wheelbarrow operation to its current size. His lost fingers were part of his legend and he was not embarrassed. To him they were like scars from years of combat with the devilish mine. He considered the stubby remnants proof of his commitment. "Like medals," he would say.

"Our morale was down, and now and then the guys would cuss each other out. We just wanted to leave," said miner Pablo Rojas, a third-generation miner known as a hard worker with few words. "Each man had his own personality." And many of the men had their own addictions. In addition to tobacco, many had a prodigious addiction to alcohol. For these men, entrapment also meant enforced cold turkey, the accompanying mood swings and desperation making their ordeal all the worse.

Three days had now passed. The men had been trapped for seventy-two hours. This was far longer than any of them had ever spent underground. Despite their efforts to attract attention, there had still been no contact with any rescue team. Food was scarce, the water terrible. The eerie echo of cracking, shifting rock was followed by silence — a reminder that they were deep inside the belly of a beast, swallowed and trapped far below civilization.

The miners grew desperate. While they tried to avoid the question, a singular reality began to haunt them all: Will we get out of here alive?

FOUR
SPEED VS. PRECISION

Day 3: Sunday, August 8

Copiapó, Chile, is a city surrounded by undeveloped beaches, vast desert and barren mountains packed with rich deposits of gold, silver and copper — worth millions, in some cases billions, of dollars. These hidden treasures were first mined in 1707, when the population of Copiapó was 990 residents. Today the city is still modest-sized — 125,000 residents including the surrounding areas — yet the local airport is literally buzzing. There are fourteen daily flights to and from Santiago, flights often fully booked as a flood of mining engineers, geologists and surveyors arrive in Copiapó. They disembark the airplane, stepping down a steep metal staircase, then wander unescorted across the tarmac — sometimes mistakenly heading to the baggage area — to a miniature terminal where vendors sell oysters, crab claws and *locos* (abalone), a delicious local shellfish with firm white meat and a delicate taste like lobster.

These executives are forward scouts for an army of entrepreneurs seeking to harvest the profits from the most recent worldwide copper boom, which began in 2002 and by August 2010 showed no signs of collapsing. With Chinese industry maintaining an apparently insatiable appetite for copper and minerals, mining operations in Chile continue to boom. Every day Chile exports approximately $70 million worth of copper, and every few months a new multimillion-dollar mining project is announced. This region of northern Chile is also home to one of the world's great concentrations of high-tech mining equipment — machines capable of pounding, drilling and grinding through thousands of feet of solid rock.

Four days after the collapse of the San José mine, Chilean President Sebastián Piñera mobilized the vast fleet of mining machinery like a rogue army general. Rejecting advice from aides to be cautious, Piñera bet everything that he could save the miners, putting his face forward as a guarantee for a successful mission, thus provoking panic throughout his inner circle of advisers, who thought the president had just volunteered for a kamikaze mission at a workplace where even the miners who dared work there gave themselves the same brave moniker.

Among Piñera's first assignments: find a general manager for the rescue operation. Long a fan of hiring multilingual MBA whiz kids as his personal inner circle, Piñera was out of his element in the world of mining. The mining minister, Laurence Golborne, whom Piñera appointed in March 2010 at the start of his presidential term, was also an outsider to the mining world. Golborne had been appointed to Chile's top mining post in recognition of his management skills as a chief executive with Cencosud, an $11-billion-a-year chain of retail stores and upscale South American supermarkets. The leaders of the Chilean mining industry were unimpressed by the dashing executive with a penchant for blasting youthful rock music on his iPhone. His blithe answer to their concerns did little to encourage them; in response to the question of how he would overcome his inexperience, Golborne had quipped, "I'm a fast learner."

Piñera and Golborne collectively had only a superficial understanding of the mechanics of subterranean mining; their knowledge of how to organize a rescue of men trapped underground was even more minimal. The men turned to Codelco, the state-run mining conglomerate that produces 11 percent of the world's copper supply. On August 9, after a barrage of phone calls and hastily arranged conference calls at the highest levels of Codelco and the government, Piñera found his general manager for the rescue. But no one bothered to tell the unwitting candidate.

When the call finally arrived late at night on August 9, André Sougarret was in bed, ready to sleep. "The board of directors has decided, get yourself and a team . . . to help the people who are in charge of the rescue," said his boss at Codelco. Sougarret, a calm, forty-six-year-old engineer with the hint of a smile never far away, listened attentively, but was not particularly moved by the message. He mentioned the call to his wife, then went to sleep at his home in Rancagua, Chile.

Sougarret had been in mining since he was in his mid-twenties and never failed to make friends as he worked his way up the Chilean mining industry's chain of command. With a specialty in underground mines, he was currently manager of mines at El Teniente, the world's largest underground mine with 1,500 miles of tunnels and a workforce of 15,000. In 2009, this mine produced 400,000 tons of copper. If El Teniente were an independent country, it would rank twelfth in the world for copper output.

Sougarret was aware of the collapse at the San José mine but never considered it an affair of the state-run copper giant. The accident had occurred at a privately owned mine about 600 miles north — a disaster, yes, but someone else's disaster.

Day 4: Monday, August 9

At 10 AM, Sougarret received another call, an urgent order: come to the presidential palace immediately. "I thought, this must be a mistake," said Sougarret. "Why would they call *me* to La Moneda — the presidential palace?" Sougarret packed a tiny knapsack, grabbed his miner's helmet and drove ninety minutes to La Moneda. He had passed by the building hundreds of times but never entered it. Ushered to the second floor — home to the office of the president and his top strategists — Sougarret was told nothing, just asked to wait.

La Moneda is pockmarked with history and had he looked, Sougarret would have noticed the walls speckled with hundreds of now-patched bullet holes, a lingering testament to the September 11, 1973, military coup d'état, which blasted then-president Salvador Allende from this seat of power. Allende, an aristocratic physician with a deep allegiance to his socialist revolution, resisted the army attack, firing back from a second-story window with a machine gun, allegedly a gift from Fidel Castro. Allende's body was found after the siege, with a single bullet hole in his head. Most historians agree it was suicide. For the next seventeen years General Augusto Pinochet ran Chile with a combination of Spanish Inquisition torture techniques and highly modern economic reform. Three thousand Chileans died at the hands of the military, but the steady economic growth established Chile as Latin America's most stable economy — a juxtaposition that for the subsequent decades spawned zealous foes and fanatical fans of the now-deceased general.

Following Pinochet's rule, bloody memories of torture and execution convinced a generation of Chileans to boycott right-wing politicians. From 1990 to 2009, the country was run by a series of progressive presidents who attacked poverty, invested in infrastructure, promoted personal freedoms and signed free trade agreements with dozens of nations. The 2010 election of Piñera — a centrist politician from Renovación Nacional, a right-wing party — buried the ghost of Pinochet and ushered in a new kind of government: technocrats with something to prove. The Piñera inner circle knew that being seen as right wing in Chile meant being permanently on probation — that if they failed to lead Chile, it might well be another generation before any of them received a second chance.

Inside La Moneda, Sougarret felt uncomfortable. He had dressed informally, in blue jeans. His miner's helmet and knapsack were in contrast to the whirl of smartly dressed suit-and-tie-clad men. A phalanx of journalists hovered in the halls, confirmation that something urgent was happening. Yet Sougarret was ever more confused; in two hours hardly a word had been spoken to him.

Finally the message came — "Let's go!" — and Sougarret was whisked to the basement garage where he boarded the presidential motorcade. Flanked by cars, each with Uzi-toting bodyguards, Sougarret crossed Santiago in a rush of protocol. Entering the airport, the convoy ignored the commercial gates and headed to Air Force Group 10, home base for the presidential jet. Sougarret still had had no briefing, no word on his mission or destination. Aboard the plane, President Piñera summoned Sougarret to his private cabin, pulled out a sketch pad and made a crude drawing of the mine and the safety shelter and issued a directive: get them out. Piñera told the still-baffled engineer to design the best possible rescue plan. Piñera emphasized that the operation was assured the full backing and resources of the government.

Only then did Sougarret realize he had been drafted to lead the mission. Thirty-three lives were in his hands and no one had asked if he was available, willing or felt capable. Sougarret would later compare the experience to being kidnapped.

Arriving at the darkened camp, Sougarret was further disoriented. He had never visited the San José mine, and again, without warning, his responsibilities grew. President Piñera announced to the gathered media that he had brought an "expert" who would take responsibility for the rescue.

"Okay, I thought, this is getting complicated," said Sougarret. "We then walked some steps toward the campsite where the family members were. I was struck by those anguished faces. . . . There were fifty people. I noticed many worried faces and, in some cases, desperation. And unease. I remember they said some nasty things to the president, because he had first gone to the press and then to them. That was a pledge we always fulfilled — to speak first with the families and then with the journalists. That stuck in my mind. Then the president explained that he came with these experts who would try and solve the problem, and would utilize all the possible resources. That was a key moment for me, the beginning of it all," said Sougarret in an interview with the Chilean newspaper *El Mercurio*. "I realized that I was in charge of the operation. The president left and I was there, alone."

Sougarret didn't need distressed family members to hammer home the consequences of a mining disaster. El Teniente, where Sougarret was a top manager, was the site of Chile's deadliest mining accident, known as the "Tragedia del Humo." The 1945 "Smoke Tragedy" at El Teniente was sparked by a fire inside a storage bunker. Barrels of burning oil quickly trapped more than a thousand miners behind a cloud of impenetrable, thick smoke. The smoke filled the cracks and corners of Tunnel C. For hours the miners held wet cloths to their faces, a crude measure that soon proved ineffective as the men collapsed. The mine's safety systems were substandard; emergency exits were not clearly marked.

As clouds of billowing black smoke poured from the mine, a courageous rescue effort was unleashed. Miners rushed into the flames, trudged into the inferno, and carried semiconscous colleagues to the surface. Six hundred men were saved, but 355 perished.

The Smoke Tragedy provoked a national debate over mine safety and led to the creation of the Department of Mining Security. The concept of risk prevention was introduced in management decision making and the plans were so successfully implemented that El Teniente would win international safety awards for fourteen consecutive years. It was no coincidence that aides to President Piñera, realizing that the owners of San José were unprepared for a sophisticated rescue operation, mobilized the most professional, safety-conscious team in Chile — the crew from El Teniente, with Sougarret in command.

Sougarret's first challenge was to coordinate the drilling. Throughout the four days since the collapse, the Chilean mining community had mobilized and sent convoys of equipment to the scene — heavy-duty bulldozers, water trucks, cranes and drilling machines, which could bore through hundreds of feet of rock, creating shafts known as "boreholes." The engineers at the site had quickly decided that a rescue via the mouth of the mine was extremely dangerous and that the boreholes offered the most reliable option for making contact with the trapped men.

The machines used to drill the boreholes were bunched together, a cluster of steaming, hissing towers that looked like an oil field operation interspersed with flopping Chilean flags. These portable drilling contraptions were not modern: since the 1950s, they have been hauled, shipped and portaged to random corners of the planet to perforate the outermost levels of the Earth's crust.

These machines helped locate the raw materials used to fuel a half century of industrial frenzy — everything from aquifers to zinc deposits. The drills were now joined in a communal prospecting mission, a daring search-and-rescue operation. A 3 1/2-inch-wide drill bit was aimed toward one of the spiraling tunnels nearly half a mile below: a needle with a 2,300-foot journey, in search of a mine shaft.

By the time Sougarret arrived, six different drills had been tried in haste — and chaos. "They were making various holes but there was no strategy," said Sougarret of the drilling scenario he inherited. "We decided that there would be three drilling techniques . . . three parallel plans with different concepts: some would be fast and others precise. We were fighting time. If we wanted to be precise, we would be much slower. If we wanted to advance quickly, things could deviate." Sougarret calculated that the boreholes could advance 325 feet per day; to achieve the precision necessary for the drills to be even close to the trapped men would take — at a minimum — a full week.

With sketchy maps of the mine, the drillers were forced to estimate the exact location of the underground safety shelter — the presumed spot where any surviving miners might be holed up. Could the men have made it to the shelter? The vehicle workshop? Or were they buried in the rubble? Like the rescuers who searched the shafts manually, the amount of guesswork was a frightening variable for engineers accustomed to precise instructions.

Typically a drilling operation is aimed at a massive pool of oil or an underground aquifer. Here the target was minuscule: the safety shelter was the size of a backyard swimming pool. If engineers misdirected the borehole heading by just two inches, by the time it had tunneled the 2,300 feet down to the level of the safety shelter, it would miss by hundreds of feet. "When we arrived at the mine, we were told to place our drill right above the refuge," said Eduardo Hurtado, a drilling supervisor with Terraservice, a

Chilean company that specializes in perforations. "The bulldozer began to flatten the area to make the platform for the drilling rig and the topographer was making measurements. Then José Toro, a geologist from the El Salvador mine, came. He knew the area well and told us to move the machine; they were worried that the entire mountain might still come down."

Day 5: Tuesday, August 10

Despite the huge challenge of finding the men, the sound of drilling was taken by everyone at Camp Hope as a positive sign. If any of the men were alive, the distant rumbling would alert them that rescue efforts had commenced. "I know that they feel every thud from the drills that are perforating the rock, bringing to them that which is most fundamental — oxygen, food and water," said President Piñera. "I hope that the six drilling machines that are working tirelessly will permit a happy ending. But I also want to acknowledge that this is not easy; the situation is very complex, and the mine keeps collapsing, as it has a geological fault. As the miners say, the mine is alive and that makes the rescue tremendously difficult." As the president and his team calculated the intricacies of the rescue plan, hundreds of local residents flooded into the mine site.

August 10 is Dia del Minero — Day of the Miner — in Chile, and family members, friends and colleagues of the trapped men gathered at the San José mine to prepare a grim ceremony. Under normal circumstances, the Day of the Miner is a festive tradition that includes communal *asados* (a picnic with slabs of grilled meat), dances, religious blessings and recognition of the profession that catapulted Chile into the world economy and has helped maintain the nation's relative wealth.

In 2010, all the parties were canceled. At the San José mine, a somber procession was organized. An estimated two thousand people formed a brief and painful pilgrimage. Chilean National Television (TVN) broadcast a live feed from the remote region. The nation watched as the family members slowly marched by, tears streaming down their faces. A crew of men struggled as they carried a statue of San Lorenzo, the patron saint of miners, for whom the rescue operation was named. Other men lofted on their shoulders the Virgin of Candelaria, a symbol of protection who was kept in a shrine at the nearby La Candelaria mine. After direct petition from the families at San José, the Virgin's return trip to La Candelaria was canceled; her powers were needed here. From an improvised altar on the back of a flatbed truck, Bishop Gaspar Quintana urged families to remain strong, chastised government officials for substandard workplace safety, and issued a direct petition for Dios to send good news about the trapped men. "Send us a signal," he pleaded. "Soon."

Local governor Ximena Matas sought to first acknowledge the collective grief, then channel it toward a team of psychologists that local government officials had cobbled together. Her team had raided the psychology department of the local anti-drug offices and imported other psychologists from Santiago to counsel the families. "We understand without a doubt that these have been very brutal days and nights for you. We have all seen how difficult it is to live this, but we have also seen the support amongst you and the strength you have shown in this wait for your loved ones," said Matas, who then outlined the importance of sharing those feelings with the mental health professionals brought in by the government.

Day 6: Wednesday, August 11

The two hundred family members awoke to yet another surprise in the desert: heavy rain.

An unusual storm hit the Atacama, turning the dust-clogged tent city into a bitterly cold and slippery mud patch. Some families spent the day inside cars, the heat on, windows rolled up. "We are not going to abandon my brother," said Jeanette Vega, whose brother Alex was trapped. "As long as he is not out, no one moves from here despite the rain, the cold or the sun." Throughout the day, family members huddled in sleeping bags, drank tea and coffee from Thermoses, and used clear plastic tarps to shield the tents from the rain. Army troops came in and began erecting a flat, more sheltered area for a series of tents — an area where family members would be more isolated from the constant stream of trucks and vehicles. By evening, the temperature had dropped to 28 degrees Fahrenheit and the families clustered around bonfires. Local vineyards had donated stalks pruned from aged grapevines; the hard, twisted wood burned slowly and hot — and the blazes continued until daybreak. Sleep was no longer part of daily life.

Day 7: Thursday, August 12

A week had gone by with no signs of life from the miners. Rescue workers had seen their options narrow as the mine continually shifted, crushing ventilation shafts and burying hopes of an imminent rescue. The Chilean mining community rushed to send rescue teams, high-tech search tools, and as many drills as necessary, but from the missing men there was not a word.

"This is not the moment to be demoralized. My colleagues can probably hear the sound of the drilling, and that will give them more strength," said Gino Cortés, the worker at the San José mine who, five weeks earlier, lost his leg when it was sliced off by falling rock inside the same mine. "These are tough miners," Cortés said. "We know the dangers we are exposed to so there is a certain level of acceptance."

At 2 PM (the estimated time of the collapse a week prior) sirens and horns were blasted and church bells rang but the immediate silence afterward was more telling. A week and still no news. Family members continued to migrate up the mountainside, decorating the hillside with photographs and painting rocks with the names of the loved and lost. They settled in, pitching tents and beseeching rescue workers to never give up on the buried men. Camp Hope became a teeming, living shrine.

The camp evoked a sense of instant community — which was hardly surprising, as the majority of the miners were local residents, neighbors, cousins and, in the case of Renán and Florencio Ávalos, brothers. These miners were the sons of frontier-style families with six, eight, ten children, often with multiple fathers. Trapped miners Jorge Galleguillos and Darío Segovia each came from a family with thirteen siblings. With an average of eight siblings apiece, these miners came from families twice or even three times the size of the average Chilean family, a reality now reflected in the burgeoning numbers at the camp.

In response to the mushrooming population of Camp Hope, the Chilean Army mobilized to deliver portable toilets and food. A field kitchen was established, with meals scheduled four times a day — the fourth being *once,* Chile's 6 PM version of British teatime. Ivan Viveros Aranas, a Chilean policeman working at Camp Hope, was heartened by the outpouring of help. "The country has shown a unity independent of religion or social class," said Aranas, who no longer patrolled the camp but spent most of the time in conversation with the family members or playing football with the growing horde of young children at the mine site. "You see people arriving here just to volunteer; they have no relation to the trapped miners."

As Sougarret's operation expanded, the hillside around the mouth of the San José mine filled with makeshift structures: shipping containers were converted into offices, crude awnings were thrown up to shade engineers from the blistering desert sun, a pair of motor homes now housed the tiny but expanding crew of on-site journalists. Dozens of helmeted men in reflective jumpsuits awaited instructions. A caravan of fire trucks, bulldozers and ambulances arrived on site. Tractor-trailer trucks slogged up the final curves, an always-heralded parade of high-tech equipment responding to disparate teams of engineers scurrying to design a solution. Despite the grim prognosis, the camp was alive with the spirit of survival, and every few hours a distant honking announced the arrival of another load of free supplies or rescue equipment.

A dull rumble became the backbeat at Campamento Esperanza. Like traffic in a metropolis, the mechanical groan of diesel engines eventually begins to sound natural, akin to wind or the tides. Above the camp, a brilliant sky offered thousands of sparks and stars, irrefutable evidence that the world's top astronomers chose well when they invested billions in observatories in this corner of South America. Students of the stars consider the Atacama region the planet's best lens for exploring other worlds. For the residents of Camp Hope, their prayers went skyward but their every worry was directed deep down, into the earth.

Minister Laurence Golborne was finding sleep an elusive companion. The rookie politician responded to a flurry of orders from President Piñera, abandoning his wife and six children for a single obsession: the fate of the thirty-three. But a week into the rescue efforts, Golborne was distraught. Internal studies commissioned by the government were pessimistic; one study summed up the probability of the men being alive at 2 percent. Desperate, Golborne secretly went to hear a psychic. She told the minister that she had a vision of the men: sixteen were alive and one had a badly injured leg. Afterward Golborne was unsure what was more incredible, the possibility that the men were alive or the fact that a government minister had paid attention to a soothsayer.

Proposals flooded into Golborne's office — along with ideas, donations and theories. One company proposed using the borehole to deliver one thousand rats into the caverns. Each rat would have a panic button strapped to its back. Once released, the rats would scatter and scamper through the innards of the mountain, allowing a trapped miner to first catch a rodent and then press the button. Hearing the rat alarm, rescuers would then be assured that survivors existed.

On August 12, Golborne went public with his doubts. "The chances of finding them alive are low." Following his comments, Golborne was inundated by a tidal wave of criticism. Family members were devastated; keeping the faith was a tenet of their emotional stability. Doubts were akin to treason. A day later President Piñera was forced to step into the fray. He announced, "The hopes of the government are more alive than ever." Piñera's optimism stemmed from insider knowledge: the drilling was advancing twice as fast as expected. In less than two days, one of the drills had reached nearly 1,000 feet. At that rate, contact could be made forty-eight hours later, by August 14. Yet even if the borehole arrived at the workshop or tunnel or refuge, the hole would be no bigger than an apple, and then what?

Day 8: Friday, August 13

Engineers at Codelco scoured the world for technologies to improve the drilling and, if the miners were found alive, provide food and medicine to the distant men. The solution came from a local physics professor,

Miguel Fortt, from the Universidad del Mar, in Copiapó. On twelve previous rescue efforts in Chile and abroad, Fortt had been on the front line of life-and-death mining disasters. As a former miner, Fortt combined the practical knowledge of life inside a mine with the technical experience gained from varied rescue operations. Using 10-foot tubes of PVC piping, Fortt envisioned a system by which these tubes could be stuffed with bottled water and food and then lowered by rope to the depths of the mine. It was an audacious and optimistic plan. Testing began immediately. The Chilean invention was named "*paloma mensajero*" — the messenger pigeon — later shortened simply to "*la paloma.*" Fortt's ingenuity would soon be displayed to a worldwide audience.

As the government scurried to round up the tools and professionals needed for the complex rescue, two voices were notably absent: the mine owners. From the day of the collapse, when they failed to alert the families promptly, to their inability to provide accurate maps, the attitude of Marcelo Kemeny and Alejandro Bohn was widely criticized.

To family members the refusal by Kemeny and Bohn to take responsibility for relaying news of the accident was criminal. Jail time was considered too lenient. According to relatives at Camp Hope, the real way to punish the owner of a dangerous mine was to sentence him to "mine jail." It was rumored that in China, a mine operator found guilty of negligence was sentenced to serve time underground, inside a mine, fixing the very safety problems that had led workers to their deaths.

Even before the accident, the owners of San José mine owed the Chilean government more than $2 million. The entire company was financially unstable, tottering under piles of debt and a safety record that was equally precarious. Despite producing an estimated 6,000 pounds of copper a day (worth an estimated $22,000) and with gold reserves estimated at some 600,000 ounces (at 2010 prices worth nearly a billion dollars), the owners were clearly desperate. The 2009 annual report's concluding page read, in bold letters:

> **SITUATION**
> PROBLEMS TO GO ON WITH THE MINE.
> CLOSE THE MINE.
> SELL.

Chilean justice is usually slow and sabotaged by procedural motions filed by defense attorneys. However, the huge media attention on the San José collapse launched at least three separate investigations — the Chilean Congress, government prosecutors and a team of private lawyers working on behalf of the families. President Piñera warned: "There will be no immunity" for those responsible. He then decapitated the leadership of Sernageomin, the government's mine safety agency, firing three top officials. Documents confiscated from the offices of San Esteban Primera, the mine company that owned San José mine, as well as government inquiries immediately led to questions about how the accident-plagued mine had managed to reopen after numerous fatal accidents. Suspecting that at least some of the miners were dead, government prosecutors mapped out a trial strategy that included charges of homicide.

Day 9: Saturday, August 14

Sougarret, who feared being held responsible for a "rescue" that ended with a body count, ordered a change in technology. The two most advanced drills were stopped and the entire shaft hauled out of the ground. As he had feared, the enemy of speed was precision and the speedy holes had veered off course. With new drill parts flown in from the United States and Australia, Sougarret was confident the drills would stay on course — though at a far slower pace. Recalibrated and reconfigured with the new technologies, the drills began operating once again.

On the hill a good-natured rivalry erupted between the nine different drilling crews. The men were veterans of the same army, a band of itinerant drillers who roamed northern Chile providing service to Anglo American, Codelco and other billion-dollar ventures. The conditions at the San José mine were, given the standards of modern mining, practically pre-industrial age. As they ate together the mess tent became a kind of entrepreneurial information exchange; the drillers were aware of the historic and unique nature of their mission. What worked to find oil, gas and water was only partially useful in the search for trapped men. "When we are drilling for minerals, it is expected that the drill will go off course by up to seven percent; that is normal and almost expected," said Eduardo Hurtado, from Terraservice, the drilling company donating equipment and personnel to the search for the miners.

Day 12: Tuesday, August 17

For the first time, a probe passed the 2,000-foot mark, though it was losing power due to a major oil leak. "This drill can't be corrected; it is aimed at three tunnels," Minister Golborne told the media. "It could hit a tunnel or could pass between them. It has three chances to hit tunnels; it could hit or it could miss." Like a long-distance golf shot, the drills were launched in a smooth parabola. The goal was to pierce one of the tunnels en route. Asked about the responsibility of the mine owners, Golborne said those questions could wait. "Here, we are focused on the most important goal — making contact with the men. Our priority is to find them. We have many people volunteering, working intensely. Let's concentrate on the positive," said Golborne. "The costs and responsibilities have time. Our *compañeros* underground don't."

Day 13: Wednesday, August 18

Golborne's warnings were prescient. The drill shaft closest to the suspected location of the men passed by all three tunnels — missing each one. As distraught engineers watched, the drill kept up a smooth pace. Hitting an open cavern would have been evident — yet the bits chewed straight through rock without pause. At 2,400 feet the operation was halted. Plans for the wildly anticipated contact were canceled. Instead, despair filled the air when Sougarret informed the families. "This mine did not have the standards from which we could base engineering work," he said. "A mine has to file the monthly updates, to have blueprints of all the working areas; those were not there. That's why the blueprints did not coincide with reality; the topographic information was not precise."

For Sougarret, facing the distraught wives was wrenching. He knew the women who begged him for solutions were on the verge of becoming widows. Saving thirty-three men at 2,300 feet was unprecedented. No such rescue had ever been mounted. The daily challenges included invention, execution and improvisation. Many miners thought the drilling was unnecessary; they proposed rescuing the miners using the time-honored technique of hacking, dynamiting and forcing their way through the blocked tunnel. Government PowerPoint explanations about the size and thickness of the rock blocking the mine's mouth did little to dissuade them.

Encouraged by the impatient family members, a group of local miners — hardy men willing to swing a pickax all day — protested. "Let us try," they insisted. The families rallied behind the new plan. Frustrated by delays, near misses and growing desperation, even a half-cocked idea was worth exploring. But Golborne disagreed. The rebel miners began to join up and head toward the mouth of the mine, but a contingent of policemen blocked their path. Fearing trouble, thirty more policemen were rushed in. They brought reinforcements, including an armored truck used to launch tear gas canisters and blast high-powered jets of water from an adjustable nozzle fitted to the roof. Nicknamed "El Guanaco" after the llama-like animal that inhabits this desert and is known to spit, the vehicle was well known to all Chileans. The force of the water could fling an adult to the ground; the tear gas was capable of suffocating infants. The sides of the truck were rippled with dents and scars from previous battles.

Government officials pressured Sougarret to let the rogue group in, either openly or by turning a blind eye to a clandestine attempt. Sougarret held firm; any option that included the likelihood of a rescuer being trapped or killed was enough to make him walk off the job. One older miner approached Sougarret. He explained that his son was trapped in the mine. If he wanted to risk his life to save his son, he didn't see what was the problem.

“He came to me and asked me what I would have done, if it had been my son,” said Sougarret. “That struck me. I couldn’t get that out of my head.”

FIVE
17 DAYS OF SILENCE

Day 4: Monday, August 9

Like any platoon commander, Mario Sepúlveda knew that troop loyalty starts in the stomach, then migrates to the more conscious corners of the brain. With his *compañeros* starving, stressed and divided, Sepúlveda went scavenging. Using tuna from a can, Sepúlveda took the lid of an oil filter, flipped it over and voila! A cooking pot. With tuna and water he concocted a watery broth. It was hardly a meal, yet it gave the miners a taste of fish and the reminder of food. The men ate together, prayed, took pictures with José Henríquez's cell phone (the last one still functioning), and rested. It was a brief respite from the madness. Several of the miners would later note it as a key moment in Sepúlveda's ascent as alpha male.

After four days, the dust settled and, with their minds still spinning, the miners began to explore every nook and crag of the mine.

They searched for escape shafts and examined the water stored throughout the mine. Months had passed since the tanks had been filled, and José Ojeda, a rotund, balding man with a complicated case of diabetes, described the polluted liquid as so revolting that he began to prefer another option: drinking his own urine. "I did drink my urine. I told the others and they called me crazy," he said. "I knew that the Los Uruguayos had done the same."

When the miners cited "Los Uruguayos" they were speaking in code to avoid a direct confrontation with their deepest fear: that they would soon be forced to eat one another. In 1972, flying from Montevideo, Uruguay, to Santiago, Chile, for a rugby match, a squad of young Uruguayans miraculously survived when their plane crash-landed in a remote section of the Andes, on the border between Argentina and Chile. After days of starvation, the men began to eat their companions who had perished in the crash and a subsequent avalanche — initially gnawing on hardened chunks of flesh, later defrosting frozen corpses to cook the flesh on heated strips of metal from the fuselage. Nando Parrado and Roberto Canessa, two of the survivors, hiked for ten days through the mountains until they were found by a Chilean rancher. The story shocked the world. For Chileans, Los Uruguayos was not just a historical anomaly but a gruesome reality that happened on the border of their nation. Inside the mine, the specter of cannibalism accompanied the men from the first days of starvation.

From the first moments he was trapped in the mine, Victor Zamora, a bushy-haired man with Bob Marley on his brain and a marijuana leaf tattooed to his inner arm, was sure he had landed in hell. Never a religious man, Zamora adapted to his new world by relinquishing his fate to God, constantly cracking jokes and asking only that if this was the end, it be a peaceful finale. "Our only options were wait to be rescued or die," said Zamora.

While he worked long hours to organize daily tasks, shift foreman Urzúa was similarly passive in his acceptance of fate, telling the miners, "If they find us, good; if not, that's it." Urzúa's mild manner and soft voice did not reflect his rank, nor were they the typical characteristics of a field commander. Urzúa's style was a striking contrast to Sepúlveda's gung ho, hyper-proactive leadership.

Sepúlveda and Urzúa were soon given the highest authority: control over the rapidly dwindling food supply. Correctly deducing that they were going to be trapped for days, the men initially ate their tiny portion every twelve hours. But before the first week was over, Sepúlveda and Urzúa reduced the meals to once every twenty-four hours. Their stash of emergency food was divided into minimal portions — the spoonful of tuna fish accompanied by half a glass of milk or juice and a cracker. Gathered together, the men would wait until all thirty-three had been served, then in unison they would consume their meager "meal."

Decisions were not dictated by Urzúa, Sepúlveda or Mario Gómez — though Gómez commanded vast respect for both his sensible advice and his experience. As the days passed, the miners continued to debate and vote on decisions, arriving at a consensus or agreement after hearing opinions and searching for solutions. Sepúlveda was the unofficial moderator; asserting his voice as a constant interlocutor, he finessed his relationship with individuals and the collective group. "I held myself strong in front of the other men," said Sepúlveda, "but when they were asleep, I cried. I wished I had a magic wand to make a bed appear, or food."

With the advent of a democratic system, the miners established a rudimentary sense of order, organizing daily routines and tasks. Sepúlveda began to assign specific duties to each man. With mechanics, electricians, engineers and heavy machine operators present, Sepúlveda knew just how to exploit this wealth of knowledge. "I just said to Ariel Ticona and Pedro Cortés, 'You and you are going to be in charge of technology.' I gave all the people something to do down there. That was my idea."

Despite Sepúlveda's leadership, the miners still accorded Urzúa the respect of a superior and never dethroned the shift foreman — a reflection that order was maintained. "For a miner, their shift leader is sacred and holy. They would never think about replacing him. That is carved in stone; it is one of the commandments in the life of a miner," said Dr. Andrés Llarena, an official with the Chilean navy. "[Urzúa] is a leader in his field and has been for ages. He is recognized by his peers."

The men survived by following a strict regime of daily activities, including prayers and group meetings, and by keeping physical movement to a minimum unless absolutely necessary. One job that was deemed essential was to *acunar* the ceiling, which involved a group of miners prying loose rocks from the roof and letting them crash to the mine floor, thereby lessening the chances of a colleague being inadvertently "ironed" by falling rock. The more the men worked, the more civility bloomed, as they came together as a team. With basic lighting devised and strung by Edison Peña, and with headlamps charged, the men were able to simulate night and day by turning the lights off every twelve hours. Light made their existence less drastic and provided the smallest semblance of normalcy in the otherworldly environment. The lights also allowed the men to gather as a group, including at the 1 PM meeting where they made communal decisions.

Following their "town hall" at one o'clock, the men prayed. Catholics, Evangelicals and atheists united in a single vision of hope, led by José Henríquez, whom the men quickly dubbed "the Pastor." Henríquez's evangelical sermons were jotted down by Victor Segovia, the designated "chronicler" of the men's daily duties and their epic challenges. "I was the operator of a bulldozer, and inside the driver's compartment I had paper and pen, which stayed dry. That is why I became the writer," explained Segovia, who years earlier was nearly crushed by a block of rock. That accident left him in a body cast for weeks. Now Segovia's notebook became a journal of the miners' daily activities, like a ship's log.

"These men were trapped in their 'office'; they were not tourists who went cave visiting. They know the drill, know how to get around," said Dr. Llarena. "They regularly spent ten to twelve hours down there in the heat and humidity, and that's what they're doing now. It is a long shift, a very long shift, but still a shift."

Miguel Fortt, a Chilean with vast experience in mine disaster rescues, emphasized that the miners were organized as a team before the collapse. "It is similar to a shipwreck," said Fortt. "The miners had to organize in a way that promotes the survival of the maximum number of people; that is something we have in our genetics. The survival instinct is incredibly strong."

With a vast supply of water and limited but sufficient air, the miners' primary concern was food. The minimal daily rations — roughly estimated at 25 calories of tuna and 75 calories of milk — meant the men were on an unsustainable diet. Given the practically unlimited supplies of water, the men had a life expectancy of four to six weeks — possibly less as humans often die first from infections, which tend to take advantage of the body's weakened state. The nonstop heat forced their bodies to burn calories in an attempt to stay cool, while draining the body of electrolytes via the constant stream of sweat.

Many of the miners were overweight — an advantage when the body is forced to convert each pound of fat into an estimated 3,600 calories. The chunkier men were like seals in a lean hunting season: their bodies harvested the stored fat. But the initial days were brutally uncomfortable due to hunger pangs that racked their stomachs. For the thinner men, the process of converting fat to calories soon advanced to the next best source of energy: muscle mass.

As their muscles shriveled, the men noticed unusual growths on their bodies, and stained patches of skin began to form on their chest and feet. The sweltering heat and constant humidity proved an ideal medium for a powerful breed of mold that germinated, then spread, along their bodies. Canker sores and open flesh began to fester inside the men's mouths, an indication that this environment — so unbearable for humans — was ideal for infections.

Yonni "Chico Yonni" Barrios became the group's de facto doctor. A small, frail-looking man who had spent years browsing medical texts and painting with watercolors, Barrios wasn't supposed to be in the mine on the day of the collapse. He had finished his seven-day shift and was scheduled to have a day off, but he switched when he was offered double his daily wage to continue working. Inside the mine, Barrios had little time to lament his luck; he was constantly being consulted about ailments.

"He always wanted to be a doctor. He reads so much and he really knows everything about medicine," said Marta Salinas, his wife. "He would give injections to his mother and was constantly reading." As he examined the men, made recommendations and tried to keep their spirits up, the miners gave Barrios a new nickname. Inside the mine he was known as "Dr. House."

Not all the men were ready for steady tasks. After spending the first day of entrapment recovering from a gruesome hangover, Pablo Rojas was now busy fighting the demons in his head. Rojas, a round-faced man with an easygoing demeanor that emanated tranquillity, had nursed his ailing, alcoholic father for a decade, until a week before the collapse, when his father died. Not only was he still mourning his father — a lifelong miner — but all the postmortem paperwork was still unfinished. Nothing could shake the image of his father from his head. For Rojas, being trapped in the mine was torture.

Rojas scoured the cave for anything to eat. "There were no bugs or rats in the mine," he said. "Otherwise we would have eaten them, without a doubt." Rojas had never felt safe in the mine. He always sensed an imminent tragedy. In 2005 Rojas quit his job at the San José mine when those worries peaked. But he returned in 2010, lured back by the high salary. Now he was indignant — with the owners, with the mine and with himself. How had this happened? When his much-anticipated accident occurred, how had he been dumb enough to be inside the mine?

Day 5: Tuesday, August 10

On the fifth day of their entrapment a faint rumble sent vibrations down to the men.

The distant echo was an unmistakable sound that every miner recognized: a drill was coming toward them. Some men later wrote that it was on August 8, Day 3 of their entrapment, that they heard the sound; others insist it was August 9. With almost all reference points, including the sun and stars, obliterated by nearly half a mile of solid rock, the miners' recollections of timing are less than exact — and hardly significant compared to the unanimous sensation of hope inspired by the distant drilling.

Alex Vega held a piece of hollow bamboo to the wall, amplifying the sound and providing clear evidence that a drill was headed straight toward them. Vega's enthusiasm sagged, however, when he soon discovered that from any portion of the nearly 1.25 miles of tunnels, the bamboo to the wall provided a similar sensation of proximity. Only two of the men had ever worked with borehole drilling machines, and they both knew the process was fraught with failure. "I told them that the first fifty meters [165 feet] would be fast, but after that, the drilling slows down," said Jorge Galleguillos, who along with José Ojeda knew firsthand the procedures for long-distance drilling.

With Galleguillos's words of caution reverberating as loud as the drill itself, the drilling sound became both heartening and frightening. A rescue attempt had begun, yet the sound was so faint and distant, the men realized that at 2,300 feet deep, any tunnel would take weeks to drill and extreme precision to find them. Even in soft rock, the machines rarely advanced more than 250 feet a day, and the miners were all aware that this mountain was packed with some of the hardest rock they had ever encountered — rock twice as hard as granite. The men were briefly enthused, but hunger and fear were not assuaged by a drill that for all practical purposes sounded like it was still on another planet.

At night, several of the men would jump up from their beds and begin screaming at the drills. "Come on, you fuckers, when are you going to break through? Damn those assholes!" The men would fall back asleep, but wake up two hours later to again hurl curses at the walls.

On Day 9, food rations were reduced again. From every twenty-four hours, the men decided to eat just once every thirty-six hours — a tiny ingestion of food that did little to trick the body into thinking it was being fed. Starvation and fatigue reduced the miners to minimal movements. They spent the day sleeping on cardboard, conserving what little energy remained. Food was so scarce the men's small intestines shriveled.

"God gave me the strength to combat the anxiety and hunger we suffered," Raúl Bustos later wrote in a letter to his wife, Carolina. "Down here we almost fainted. I prayed and asked for us all, if death happened, that we would take it well."

■ ■ ■ ■

On the eleventh day, Sepúlveda collapsed. The pressures and stress of the extraordinary responsibilities he had hoisted upon himself were too much. He cried. He lay in bed. The captain of the mission was now himself collapsed on his motley bed.

The other men rushed to help him. Bringing Sepúlveda back was key to the group's survival.

"You can't go, Mario. Without you we aren't going to make it," Victor Zamora pleaded.

"We were like a family," said Samuel Ávalos. "When someone falls, you pick them up. But he was giving up. He simply collapsed, threw in the towel. As a group we understood the pressure he was under, but we also made him understand that he could not abandon the boat. We had given him this leadership."

The group resuscitated him. Zamora told him jokes. Ávalos began to take long walks with Sepúlveda. "I told him, 'Don't fuck with us, Perry. We have to get out of here.' "

As Sepúlveda came back to life, the group coalesced. More than ever they appreciated their eccentric leader. Alex Vega said, "Mario. Even with his madness, he saved us."

The thirty-three miners trapped in a collapsed mine became the unwilling subjects of a cruel test, a unique psychological challenge experienced by few humans. Cut off from the world, they lived in a tunnel with no natural light and — barring the gurgle of water — no natural sounds. Instead they were subjected to an unpredictable but ongoing soundtrack that included the shrieks, groans and fracture of rock. Like the mine itself, the men lived under tremendous stress.

"What happened down in the mine is a lot of things which, put together, amounted to torture. They were trapped underground — that's one; in the dark — that's another; no food — another; bad water. . . . You're piling on these things which individually are insignificant but put together you have this recipe for potentially a psychological breakdown," said Dominic Streatfeild, author of *Brainwash,* an extensive study of mind control experiments and interrogation techniques. "The gold standard for an interrogator is uncertainty, fear of impending death, loss of time, sensory deprivation, no routine. These things unhinge human beings and remove their beliefs, and a lot of them were present in the mine."

The youngest of Los 33 — the appellation the men gave themselves — was nineteen-year-old Jimmy Sánchez, who began to hallucinate and suffer from nightmares. He imagined the ghosts of dead miners haunting the caverns. Hallucinations are so frequent among solo sailors, lost explorers and lone fishermen that they become enshrined as legends or myths. The luscious vision of a mermaid at sea is a fantastic solution to a deep desire. The ghosts of dead miners may well have been part of that same fragile mind-set. As the heat drained the water and energy from their bodies, many of the men began looking for God.

Mario Sepúlveda had a conversation with the Devil. "I would go to pray in a place that was very isolated, the same place where Gino Cortés lost his leg. In one of those prayers, I was praying very loudly and a huge rock fell next to me. I knew it was not God, but that it was the Devil. He was coming for me. All the hair on my body stood up." Sepúlveda began to scream at the rock, "How much longer will it take you to understand? You too are a son of God, be humble." After that confrontation, the Devil left Sepúlveda in peace.

Throughout the mine, the men saw shadows, figures and beings that would later melt away. They called these apparitions *mineros chicos,* or "little miners." "There are a lot of paranormal things in that mine," said Sepúlveda with the conviction of a true believer. Instead of calling themselves the thirty-three men, they started referring to thirty-four: God was with them; he was the thirty-fourth miner. Even the nonbelievers began to pray.

Victor Zamora began describing luscious meals he could only dream of eating — steaks with tomato and a beer. Only Alex Vega sat back somewhat comfortably, one of few with a bed, as he had dismantled the seats from a truck and converted them into one of the cave's finest sleeping devices. Another man made a set of dominoes by cutting up the safety triangle he found in one of the vehicles. The miners gathered in small groups as they confessed fears and shared dreams and took long walks, like couples, into the dark.

Ariel Ticona suffered in private agony. He was about to become a father. His first daughter, Carolina, was nearly due, or was she already born? Had it been a healthy delivery, and how was her mother? While most of the men lived for the day the rescue would finally haul them out of the hole, Ariel Ticona lived on a different calendar. Ticona's life was reduced to a single date: September 20, his daughter Carolina's due date. "I was fine going fifteen days without food; with enough water you can fill your stomach," he said. "I was prepared for another month, to keep going."

"I gave myself up to die," said Richard Villarroel, who was less than a month away from becoming a father for the first time as well. "I lost twenty-eight pounds and was afraid of not meeting my baby. We were so skinny. . . . I looked around and saw my *compañeros,* who looked so bad; that made me very scared."

Still, Villarroel battled to save his buddies. "I don't know where I got the strength. My head was fine. But I stood up from my bed and was spinning. Very dizzy. I was rocking back and forth . . . But I would go down to level 90, find some scraps of tubing, then hike up to the other levels and put oil on the scraps and burn them trying to send smoke signals."

In the higher reaches of the mine, Villarroel could see crude messages on the walls that read "Los 33" and had an orange arrow with the word "Refuge." In the first desperate days, when the men imagined teams of rescue workers arriving, they had used spray paint to indicate the position of the refuge. Now the messages looked like a hieroglyph from an earlier era.

As they lay on the floor, talking incessantly, dying slowly, Sepúlveda began to notice that the men were living a collective dream, a utopian vision of how they would live if God and the drillers collaborated to provide a second chance at life. The sound of the drill was ever closer, but their delirium making the dream difficult to sustain.

"We had lots of good times, jokes, lots of happiness," said Sepúlveda. "At one moment we said, 'When we get out of here, they are going to invite us on an airplane trip. The plane will crash and we will all survive — the thirty-three miners surviving again.' We always laughed about that."

United with his fellow miners as victims of a profound injustice, Franklin Lobos, the soccer player, began talking of forming a nonprofit foundation, a tangible expression of their collective vision, in which they could pool their earnings and promote the concepts of decent wages and livable working conditions for workers around the planet. They dreamed of a never-ending pact, the 33 Musketeers — one for all and all for one.

The phenomenon of extreme situations

provoking positive, life-changing attitudes is long recognized by religious masters who deliberately fast. For the trapped miners, collective dreams of peace and unity came easily, but the reality was much more fragile. Family ties united various miner cliques, as uncles, cousins and brothers were bound by blood and a family tradition of mining. Twenty-five of the men lived within two hours of the mine, so they shared a common language of the desert — a tough, survivor's coda that in its vocabulary, accent and values created a cultural moat to outsiders.

While the thirty-three met collectively for decision making, prayer and food, a subgroup of five miners — all subcontractors not officially employed by the mine owners — were shunted to the sidelines. "Treated like second-class citizens," said one government official. The veteran miners had little in common, either culturally or colloquially, with the recent arrivals.

Although the refuge was a secure spot for sleeping, it was also blistering hot and smelled like a locker room overrun with weeks-old dirty towels. The smell was so unbearable that Omar Reygadas started up a heavy earth mover and began to demolish the main door. What once protected the men from the dust and the dirt was no longer needed. Reygadas knocked down the entire front wall and dumped the debris farther down the tunnel. The refuge still overflowed with the smells of ten dirty, sweating men, but now an occasional current of air made it habitable. For Franklin Lobos, the air became a sweet perfume.

Day 14: Thursday, August 19

Each sleeping area began to devise autonomous rules and rules for communal living. However, in moments of crisis, those differences were surmounted by the even stronger survival instinct. By Day 14, the miners were sure that the drilling was going to reach them — but soon enough? The men devised an intricate reaction plan: when the drill was about to break through the roof, they would disperse to all corners of the tunnel, each man with a handwritten note and clear instructions to attach the letter to the drill bit. The miners were armed with cans of orange spray paint — usually employed by topographers — to paint the drill shaft to alert the rescue team that somewhere deep below, trapped like animals, at least one man was alive. Heavy equipment was readied. The miners were prepared to use the perforating machine, known as a *jumbo,* to widen the tunnel to reach the drill shaft, if necessary.

A bulldozer-like vehicle known as a "scoop" was ready to clear debris.

As the drill bit inched closer, the enthusiasm surged inside the mine.

The men loved hearing the sound of the drill. For twenty-four hours now they had spoken excitedly about their notes and plans to alert the rescuers to their survival. The men could feel the percussion hammering, just above them. Salvation had arrived. Then a nervous realization rippled through the group.

The drilling continued, but now it was below them. The drill had burrowed 2,300 feet directly to the men, and missed them. Rushing to a lower level, the men relived the anticipation and the desperation. At 80 feet below the men, the drilling stopped. Above ground and below, the silence was deafening. The men panicked when the loudest of several drills suddenly stopped. Silence was terrifying. Edison Peña began screaming that they were all going to die. José Henríquez told the men to trust in God.

"The guys began to lose the notion of time; desperation set in. They did nothing but sleep, guys like Claudio Yañez. I started to feel that seventy percent of the men were infected with that feeling, I cried and cried, but never let them see me. The circle was about to close. The circle of death," said Samuel Ávalos. "It broke me to see Richard Villarroel — his wife was pregnant. Osmán Araya had young kids. I thought that while I had one young child, at least the others were older. I was imagining that I would never see the topside again. I was more worried about my companions. They had little babies, pregnant wives. That broke me. . . . To see my *compañeros* cry and cry. That was fucking tough. Anybody would crack watching that, anyone."

"That was the darkest moment, when we went to the lowest level of the mine and could feel that the drill had gone by," said Alex Vega. "Many men decided to die. They began to write goodbye letters. Victor Zamora was first, then Victor Segovia and Mario Sepúlveda."

"We were in death's waiting room; I waited for death and was tranquil. I knew that any moment the lights would go out and it would be a dignified death," said Mario Sepúlveda. "I prepared my helmet, my things, rolled up my belt and arranged my boots. I wanted to die a miner. If they found me, they would find me with dignity, my head held high."

For Claudio Yañez, the thought of imminent death held no such peace. For days his companions had been hinting that it was time to take drastic measures, time to eat the skinny newcomer, Yañez, who had been in the mine for just three days at the time of the collapse. At times, Yañez felt they were joking, but never enough to scrape away the meaty slab of truth: the first man to die was likely to be slowly cooked and converted into food for the rest.

Daniel Sanderson, a young miner who worked inside the San José mine but was not on the fateful shift, was later a confidant of several of the miners, who wrote him letters describing the possibility of starvation. "They thought they were going to eat each other," he said.

On Day 15, the men were down to the last of their food. The preacher José Henríquez urged everyone to hold hands and pray for the two cans of tuna to duplicate. The men obliged and put their hands together on the food box. They had little to lose, and everyone agreed Henríquez had been a savior and a unifying force. Some of the men smiled and joked as they prayed for God to produce tuna fish.

On August 21, Day 16, Mario Sepúlveda was sure he would die.

Not having eaten for two days, Sepúlveda

was now vomiting the contaminated water. He wrote a final letter, offering advice to his thirteen-year-old son, Francisco: "Remember Braveheart, the warrior who protects his people. That is what you must do, take care and protect your mother, your sister . . . you are now the man of the house."

SIX
A BONANZA AT THE BOTTOM OF THE MINE

Day 16: Saturday, August 21

On a desolate, rock-strewn slope approximately half a mile above the mouth of the San José mine, Eduardo Hurtado and his six-man team drilled nonstop. From the drill site they could see the abandoned offices of the mine, a pair of simple wooden shacks that like a ghost town captured a moment of instant abandonment — drawers open, files on the desk. In the days since the accident, the floor filled with the desert dust and the open wooden window shutters flapped lazily when the wind kicked up — which wasn't often enough for Hurtado and his crew, who toiled in a broiling desert sun. The brisk wind came at night, when the sky was aglow with stars and the temperature dropped below freezing. At dawn, a tongue of thick fog lapped up the valley, drawn from the Pacific Ocean; it added another penetrating layer of cold. No one complained.

Weather was the least of their worries as they angled a drill bit toward their target, 2,300 feet below.

The drilling team ran a twenty-four-hour operation that stopped only for maintenance at 8 AM and 8 PM, to add oil and check hydraulic fluid. Theirs was one of nine operations to drill nine separate boreholes toward the trapped miners, all coordinated by André Sougarret. Each site had a seven-person team, but the approach and drills being used varied. Sougarret had gambled on different drilling technologies. Borehole 10B was powered by a technology known as "reverse air," which could drill up to 800 feet in a single day but was difficult to re-orient if it went off course. The slower but more precise technology known as diamond drilling allowed for corrections en route.

Like rays of light, the nine boreholes were angled from above, shooting in long, diagonal shafts in an attempt to enter a tunnel, the workshop or even the refuge itself. Maps of the mine had repeatedly deceived engineers, showing structures that did not exist or failing to highlight metal reinforcing rods that in a split second could decapitate a drill and erase a week's work. In a normal drilling operation to 2,300 feet, when speed was of no matter, drills routinely veered off course by 7 percent and arrived within 260 feet of the target. On this job, the entire target was a safety shelter no more than 33 feet long and just 540 square feet in total.

Time for meals was scarce, and the mess hall down below an unnecessary interruption. Every few days a box filled with one hundred sandwiches and bottled water was dropped off, donated by the owners of Santa Fe, a nearby iron mine. The rescue workers fed the scraps to a blue-and-green lizard that lived in the rocks. "Normally we would have a grill and cook meat or chicken," said Hurtado. "This was not the time for a barbecue; we had too much anguish."

Hurtado, a fifty-three-year-old with nearly two decades of drilling experience, was obsessed with time. As he fought to keep the drill functioning, the days ran together. Had it been five days since they started this latest hole? Seven days? Like the countdown clock on a bomb, every second brought the miners — if any were still alive — closer to death. For many rescue workers lower on the hill, the mechanics of drilling a 2,300-foot tube was a daunting and incomprehensible notion. Hurtado's crew had a precise understanding of the challenge; they had arrived within forty-eight hours of the mine collapse and barely slept in the ensuing weeks. "We all had an obsessive, almost violent relationship with time," said Hurtado. "I thought that if we did not arrive in the next day or two, we would find them dead."

Day 16: Inside the Mine

With only two cans of tuna fish remaining, the miners had made another painful decision. Instead of a single bite of food every two days, they stretched the rations to one bite every three days. The miners were so exhausted that even the walk to the bathroom, just 100 feet up the ramp, was now a chore. The hardy miners, men who regularly worked twelve-hour shifts, sweating, smoking and hacking away at the mountain, were now listless, the spirit of survival eaten away by the effects of starvation and a sense of abandonment. Conserving energy was an essential bodily function. Alex Vega lay on the wet, rocky slope and looked around; his *compañeros* were prone, talking but rarely standing. "We just lay down," he said. "It was too much effort to walk."

The men's health was deteriorating rapidly — on the verge of going into a free fall known as the death spiral, said Dr. Jean Romagnoli, a Chilean doctor tasked with monitoring the men's physical conditioning and nutrition. "If their health was here," he said, holding his hand high, "in another two days they would have been falling like this." Romagnoli sliced his hand down, like a guillotine. Even a simple infection that caused diarrhea was now a potential death sentence. "I can't last much longer," Victor Zamora wrote in a goodbye letter on August 21. "The only thing that I can say to my wife and children is that I am sorry."

The sound of multiple drills grinding toward the men was constant. But echoes and acoustic tricks played by the mine concealed the exact locations of the incoming shafts. What sounded like a direct buzz toward the men could be dozens of feet off course. Given the recent failures of drills that had sounded on target, then missed, optimism was muted. Still, they were on high alert. If a drill broke through, the men knew what to do — they had practiced and strategized many times. Twice before they had mobilized. Now, they wondered, would they get a last chance?

Day 16: Rescue Operation

By the afternoon of Saturday, August 21, Borehole 10B had reached 2,100 feet, less than 160 feet from the target. That the drillers felt tantalizingly close is a tribute to the dimensions of the operation. In many rescues, drilling through 160 feet of solid rock would have been a challenge, here it was just the final stage. Hurtado and his crew knew that in another twelve hours of drilling, Borehole 10B would reach the target depth. They also knew that the drill was slightly off course.

Nelson Flores, the lead operator, fought to guide the drill to a precise set of GPS coordinates derived from a software program called Vulcan, a world-class tool that uses precise digital maps and then overlays the drill's trajectory. By projecting the curvature of the drill's parabola, Vulcan allowed engineers to guide the boreholes toward the final target. Flores also asked for help from above. Every day when he arrived at the drill, he carefully opened his pocket and pulled out a rosary and hung it gently from the controls. That rosary had belonged to his sixteen-year-old daughter, who had died the year before. When he drilled, the rosary shook slightly.

Now Sougarret needed a hybrid miracle. The team led by Hurtado had kept the drill advancing rapidly and almost on track. The deviation was slight, but projections for the remaining feet indicated that the borehole might miss the target coordinates. "We didn't have much faith that it would hit. . . . We needed it to go more vertical and change direction. And that is what happened in the last stretch, when it was most difficult," said Sougarret.

Day 17: Above Ground

At 2,165 feet, Flores slowed the drill. Instead of the normal twenty revolutions per minute, he lowered it to five revolutions per minute. The goal was not to rip through the walls of a tunnel but to ever so gently puncture a clean hole. Operating at full speed, the drill bit might fire shards of rock in every direction, launching a barrage of missiles capable of wounding or killing a miner.

By 4 AM a small crowd had gathered around Borehole 10B. The night was calm, the usual wind and fog, pleasantly absent. Despite a spate of recent near misses with other drills, the anticipation was electric. Floodlights lit the scene like a movie set, casting long shadows across the adjacent rock piles. The rumble of the drill engine was interrupted by frequent pauses.

Like a divining rod homing in on water, the angle of perforation in those last feet changed ever so slightly. After two weeks of nature foiling their every effort, Sougarret and Hurtado were rewarded with a positive surprise. The drill somehow corrected its course. "In the drilling things happened that had no engineering logic. I believe that something occurred," said Sougarret as he struggled to explain the last-minute correction. Asked if he meant a miracle, Sougarret became cautious. "We had luck there . . . or help."

At 5:50 AM, as the drill passed 2,257 feet, Flores felt the entire shaft go into a brief free fall. There was no more resistance; the drill had broken into an empty space 12.5 feet deep.

Day 17: Inside the Mine

With virtually no energy remaining, the miners had long ago abandoned the idea of staying up all night waiting for a drill to arrive. Sleep had never been easy. Humid air, wet ground and the tense environment had always conspired to prevent a deep sleep. All-night domino games served to ease insomnia and combat the terror of starving to death.

At 5:50 AM, the sound of a whirring drill, crashing rock and a grinding noise shattered the calm inside the wet, slippery tunnel. "I was awake, playing dominoes," said Richard Villarroel, who was down the tunnel in the refuge. "When the drill broke through, it was the most marvelous moment for all of us. We looked at the drill and were stunned. It even took us a few moments to understand the importance of what had happened. Only then did we start to hug and celebrate. We then understood the reality. They were going to save us." Then chaos took over. "It was crazy, people were running everywhere," said Villarroel. "I looked for something to hit the tube with."

Day 17: Rescue Operation

In the dawn light at the drill site above ground, the workers jumped up and down, hugged and yelled and awaited instructions from Hurtado. Flores immediately shut off the drill.

Gathered around the shaft in silence, Gabriel Diaz, an assistant on the drilling operation, was ready with a sixteen-pound hammer. He pounded the tube three times. Immediately, Hurtado put his ear to the tube. He heard a faint rhythmic ringing from below — "like someone hitting the shaft with a spoon," said Hurtado. Moments later a series of deep metallic thuds rose from below, unmistakable signs of life.

The echoes from below were obvious; how could there be any doubt that deep below someone was whacking away at the tube?

But Hurtado and his crew were torn. A week earlier a borehole they drilled at San José had followed this same scenario — upon reaching a tunnel at a depth of 1,640 feet, the drillers heard rhythmic percussions. Then when a video camera was lowered to the depth, the men were incredulous at the sight: no sign of life. No miners. Had they imagined life below? Was the mine taunting them?

Minister Golborne and Sougarret rushed to the borehole. Like a doctor in search of a pulse, Sougarret used a stethoscope to amplify the distant thumps. Someone was pounding away. Golborne began hugging the rescue workers, then, looking stunned, he pulled off his helmet and headed down the hill, determined to be the first to inform the families. Tent by tent, the minister spread his cautious message: *We will have news today. Be alert*. The entire camp came alive with expectation. Journalists pestered the minister for answers; he remained secretive, offering only that the president would be arriving. The family members, roused from their slumber, began waving the Chilean flag and chanting, "*Viva Chile!*"

Day 17: Inside the Mine

From all directions, miners came running

to see the drill, a swarm of men with cans of spray paint, determined to paint the drill. "We were afraid it would pull up and go away. We had to work fast," said Alex Vega, who explained how the men forgot their long-rehearsed protocol. "We were supposed to first stabilize the area, to make sure the loose rocks on the ceiling were knocked away, then attach the messages, but it all happened so fast we did it backwards: everyone was working on the drill and the roof was still dangerous."

With a heavy wrench the size of a baseball bat, Villarroel began slamming the tube. A booming crack echoed inside the tunnel. But did it echo above? Villarroel switched to an iron tube from one of the mining machines. Iron striking iron, the combination sounded like a gong. The miners took turns battering the drill shaft.

With slabs of rock dangling above their heads, the men tied letters and notes to the drill shaft. Mario Gómez and José Ojeda attached their messages — to wife and to rescue officials, respectively. Other men clumsily bound their written notes to the now motionless drill. Mario Sepúlveda ripped off his underwear and tore out the elastic, which he used to wrap the messages to the drill shaft.

For an hour the men banged the tube. They sprayed until the paint was gone. Then the drill slowly rose. Again the men were alone. Now the atmosphere inside the tunnels became charged with a miraculous sense of resurrection. From the brink of starvation, cannibalism and a torturously slow death, the men were suddenly just hours away from a heavenly answer to their prayers: food.

Day 17: Rescue Operation

Back at Borehole 10B, Hurtado and his drilling crew began the arduous task of withdrawing the 114 tubes that, joined together, formed the nearly 2,300 feet of metal shaft. Disassembling the shaft, which was divided into 20-foot segments, each weighing 400 pounds, would take six hours.

As aides kept him updated throughout the morning, President Piñera was consumed with another urgent matter: his eighty-seven-year-old father-in-law, Eduardo Morel Chaigneau, was dying. Together with his wife, Cecilia Morel, Piñera was at Chaigneau's bedside, telling the dying man that the miners were sending signs of life. Air Force officials readied the president's plane. At noon, Morel stopped breathing. An hour later Piñera was rushed to the Santiago airport, and in a small plane, accompanied by his minister of the interior, Rodrigo Hinzpeter, he flew to Copiapó.

Before Piñera arrived, the last section of drill was removed. Eduardo Hurtado looked at the muddy piping and saw an orange splotch on the shaft, above the drill bit. A message? Wiping away the mud, Hurtado grabbed a gallon jug of bottled water and poured it over the drill, dousing Golborne as well. "Sorry, Minister," said Hurtado, as he cleaned the shaft to reveal a crude orange stain. "That mark is not ours," said Hurtado. "Minister, this is a sign of life."

At 2 PM, Golborne inspected the tube. Hearing a distant clanging had encouraged the minister, but here was hand-painted evidence of survivors. Seconds later, as the drill bit emerged completely, the men saw a yellow plastic bag tied to the tip of the drill. It was wound in cables and the rubber elastic from Sepúlveda's underwear. The workers unraveled the cables and peeled away layers of muddy plastic from the sodden package. Golborne opened the small shredded pieces of paper as if they were delicate gifts. He began to read aloud from pages torn out of a notebook. A message from the deep. "'The drill broke through at [level] forty-four . . . in the corner of the ceiling, on the right side . . . some water came down. We are in the shelter. . . . may God bring you light, greetings, Mario Gómez.'"

On the other side was more writing. Golborne again read aloud to the hushed crowd. "'Dear Lily, patience, I want to get out of here soon. . . .'" He continued reading in silence, and then announced, "This is personal." Golborne carefully gathered the scraps of the letter and, together with Sougarret, prepared to board a pickup and drive down the hill. Protocol weighed heavily on both men; they were determined to brief the families before the news leaked.

Francisco Poyanco, a technician on the drill rig, was stacking the metal piping coming up from the hole. The very last tube, in which Golborne had found the note, was dripping with mud and earth from below. Poyanco began to gather the nylon bags and cables that had held Gómez's note. Half buried in the mess, a lump of tape stood out. Poyanco picked it up and discovered another small tightly wrapped package — another note from the buried men. Poyanco was thrilled, thinking it was a souvenir he could take home. As he unfolded the note, however, Poyanco felt chills — "*Estamos Bien En El Refugio los 33.*" In clear red letters, evenly spaced and calmly written, was the proof of salvation: all the men were alive.

Poyanco ran toward Golborne, carrying the scrap of paper he had found in the mud. He began yelling that all of the men were alive. Hurtado heard the cries. Golborne paused, then seeing that Poyanco had a note, told him to read it aloud. The thirty-year-old assistant unfolded the note and read aloud the seven words: "*Estamos Bien En El Refugio los 33.*" ("We are all right in the shelter, the 33 of us.") The drill site erupted. Like spectators at a soccer match after a spectacular goal, helmeted engineers thrust their arms skyward, jumping up and down and hugging one another.

The reaction of Cristian Gonzalez, twenty-two, a mining technician working for his father at the San José mine, was instantaneous: he ran down the hill, into Camp Hope, screaming, "They are alive! They are alive! They sent a message that they are all fine, but they can't tell us anything!" Later, Gonzalez defended his breach of protocol. "I know these miners. I worked seven months in that mine and am close friends with Claudio Acuña and José Ojeda," he said. "I promised their families that as soon as I heard anything, I would tell them."

Day 17: Inside the Mine

With the drill gone, Zamora took charge of

reinforcing the roof. It was the same task he had completed in those last nerve-racking hours on August 5, when he had sensed that a collapse was imminent but was ordered to keep working. Zamora cleaned debris from the roof with a renewed passion — salvation depended on the integrity of this solitary hole. An earthquake could seal them off again. Every man inside the tunnel and every rescuer above understood that the miners were far from being rescued. Right now, the urgent mission was getting nutrients and medicine to the bottom of the mine.

Day 17: Rescue Operation

At 2:30 PM President Piñera arrived at Camp Hope, adding yet another level of urgency and expectation to the already frenetic scene. After a brief meeting with family members, Piñera waded into a crowd of journalists. Flanked by family members and Senator Isabel Allende — the daughter of former President Salvador Allende — Piñera held up a clear plastic bag containing the note from the trapped miner José Ojeda and read aloud the message: "*Estamos Bien En El Refugio los 33.*" "This came out today, from the gut of the mountain, the deepest part of this mine," said the president, barely able to keep his eyes open in the sharp desert sun. "It is a message from our miners that says they are alive, they are united, they are waiting to see the light of day, to hug their families."

Carolina Lobos, who had spent seventeen days sleeping with her trapped father's black-and-white Nike T-shirt, said, "I cried when I heard they were okay. . . . Everybody was yelling, 'They are alive!' 'They are alive!' I was in shock. I called my mom and said, 'Mom, they are alive! Bye.' . . . I cried from happiness. I hugged Kristian Jahn [a government official overseeing the psychologists]. He was the handkerchief for all my tears."

Camp Hope became a delirious scene of tears, smiles, hugs and waving flags. In a spontaneous charge, hundreds of family members surged up the hill to stand among the thirty-three flags that had long symbolized their faithful vigil. Each flag bore the handwritten name of a miner. Each flagpole was surrounded by a wreath of melted candle wax. As they bellowed out the Chilean national anthem, President Piñera — part of the crowd — joined in.

In minutes, the message sped throughout Chile. Strangers hugged on the subway and in the streets. The miners are alive! All of them! Drivers tooted their horns. Thousands of people flooded the streets of Santiago, heading to Plaza Italia, the usual site of soccer celebrations. It was as if the nation had won the World Cup — a joyous, patriotic uprising.

While the nation celebrated, the rescue team scrambled to outline long-term priorities. No one was satisfied with just one paloma tube down to the miners. Three separate boreholes were needed. Maybe more. An earthquake or a cave-in could quickly collapse the one fragile link they now maintained — a disaster that would send the rescue back to Day 1 and the miners to almost certain death, as the men had no reserve food supplies. Paloma 1 was quickly designated the delivery chute for food and water. The second hole would deliver enriched oxygen, water and electricity. The oxygen line was designed to pump the cavern with cold air, in an effort to lower the suffocating hot temperature. It would also deliver a permanent fiber-optic link that would allow the men to communicate with their loved ones face-to-face. The third hole was designed to come through far from the men's living quarters. This would be the hole used for an eventual escape. Though it was not clear how the men would be extracted, one theory had the rescuers first drilling a borehole and then widening it out so that it was large enough for the men to squeeze up through it. That option was deliberately separated from the other, more day-to-day functions. For the rescue shaft, the rescue team aimed for the roof of a vehicle workshop, some 1,200 feet above the men's main living quarters. It was a larger target and would provide a staging ground for the final rescue. Despite the long-term plans, everyone knew that they were far from that fantastic moment. For now the men needed medicine, food and a survival plan.

Had the miners been trapped a generation earlier, their communications would have been limited to handwritten letters and a telephone. Now engineers carefully lowered a video camera to the bottom of the shaft to gather information on the condition of the miners.

Day 17: Inside the Mine

While they waited for signs from above, the men peered into the shaft. Their lanterns illuminated a wet tunnel that quickly swallowed up the light. Beyond 30 feet they could see nothing. Water dripped down on them as they crowded around, continuously peering upward. A current of cooler air drifted down through the hole, the second welcome arrival from above. All the men were now united. Hugging and wiping away the sweat, they were already far removed from the panic and terror of the preceding days. No food had arrived, but hunger had long ago waned and disappeared. Now the men were filled with a joyous anticipation, an answer to their prayers, a renewed faith that they would have a second life.

The men began to speculate — what would be sent down first? A hot meal? Soap and shampoo? A fresh toothbrush? An instruction manual for survival? Each man began to let his imagination run free; even the ability to fantasize about simple pleasures, small treats and deliveries from above had nourished the men's collective spirit.

Three hours later a small light began to descend: a tiny object was being lowered to them. The men crowded around the hole, staring up and wondering aloud about this historic first delivery. "I thought it was a shower at first," said Pablo Rojas as he described a tube with a bulbous structure at one end. When the object popped through the roof, it was clearly a high-tech electronic device, but no one had ever seen anything like it. The mini camera was immediately lowered to the floor. Like a robotic insect, a lens cap flipped open and the camera began to rotate and rise — a remote-controlled video camera but what about the sound? Could this machine hear?

Pablo Rojas approached the camera. "What is this damn thing?" he wondered, putting his face close so he could inspect the rotating camera that was now rising off the floor. Luis Urzúa, the shift foreman, started talking to the machine: "If you can hear me, raise the camera," said Urzúa.

The men waited. The camera went down. The men laughed, giddy from a combination of adrenaline and excitement. For twenty minutes the camera whirled and recorded, then it began to rise slowly. Pablo Rojas watched the camera disappear up the shaft. "I wanted to hang on to it and have it pull me out, but I didn't fit."

Day 17: Rescue Operation Above Ground

As the world awoke to the story of the Chilean miners, back at the mine, engineers were furiously trying to fix the audio function on the video camera. The delicate machine had been damaged by contact with water. The audio was gone.

The images broadcast back to the men at the communications office were eerie and hard to decipher. Dim lights shone in the background, obviously the head lamps of miners who had crowded close to the camera. But the low light conditions made the resolution so grainy that the rescuers could only guess at the faces they were seeing. Despite the frustration at the lack of audio, the men appeared to be standing and moving about. For every question answered, a dozen more popped up: What were the injuries? Had any of the men been badly crushed? After seventeen days with minimal food, had they developed life-threatening illnesses?

Two hours later, the video was shown to family members at Camp Hope, projected onto the side of a tent. The black-and-white images were barely decipherable. At an odd angle, with only a fraction of a face visible, a pair of eyes drifted into view. The curious and haunting eyes of Florencio Ávalos. Or was it Luis Urzúa? Or Esteban Rojas? Various families claimed the eyes belonged to their lost miner. Indeed, the dark, blurry images were so generic it allowed for the instant substitution of subconscious thoughts. A Rorschach test at 2,300 feet.

With the anguish and desperation temporarily soothed by a mood of empowerment, Camp Hope became a shrine to the living.

With bonfires sparkling, music pounding, Camp Hope came alive with dancing long past midnight. At 2 AM, while family members stomped and celebrated on the rocky dance floor, volunteers handed out hard-boiled eggs, sausages and grilled chicken. Paul Vásquez, a comic nationally known as "El Flaco" ("Skinny Man"), gave a stand-up performance, while Juan Barraza, a local priest, offered a prayer session in an adjacent tent.

Barraza was encouraged by the scene. "To know that they were alive allowed everyone to express many emotions that had been held back. It was like opening a pressure cooker. Now everyone was saying, 'We won't go home without them.'"

SEVEN
CRAWLING BACK TO LIFE

Day 18: Monday, August 23

Following the excitement of first contact, the men prepared to eat. After days of jokes about juicy steaks, hallucinations of fresh *empanada* meat pies and visions of a banquet, the miners were ready for a feast. Instead, the initial doses of liquid were deliberately minuscule so as to nurse their bodies back to health. "They sent us tiny little plastic cups with glucose," said Vega, "like the amount when you give a urine sample at the doctor."

For skinny guys like Claudio Yañez, seventeen days without food had left him looking like a skeleton wrapped in a tight layer of muscles, his face gaunt.

"We expected food but it was only liquid," said Claudio Acuña as he described the men's surprise that for the first forty-eight hours, no solid food would be permitted. The men followed orders, taking their medicines and slowly ingesting the glucose and bottled water at regular intervals.

Sepúlveda lived in a bizarre limbo. His body was still crashing, the effects of starvation worsening. Emotionally he was fragile, a stew of exhilaration from first contact; anticipation of a conversation with his wife, Katty; and sheer wonder that a drill had arrived. Sepúlveda had developed such familiarity with the underground world that "the smell of mud and human skin became agreeable, part of my life." But contact from above had done nothing to ease the constant humidity. "Our clothes were wet; we walked around in our underwear," he said. At night the men slept together, side by side, on the ground.

The men readily admitted that they slept huddled and close on the tunnel floor, which brought up questions of sexual activity. The communal sleeping arrangements proved fertile turf for those who doubted that thirty-three men — regardless of their stress and suffering — could live for weeks without sex. Sepúlveda denied rumors of homosexual activity during the seventeen days of solitude, insisting that their energy level was barely sufficient for walking and talking. Sex, he said, was far from their minds.

As operator of the heavy earthmover known as a "scoop," Sepúlveda needed to activate foot pedals and consequently wore a different type of boot than the typical miner. His boots were thicker, enveloping his feet in constant humidity; this caused a severe case of fungal infection on his feet. Sepúlveda's chest and back were mottled with tiny red spots. Like a disease, the fungus spread across his body; sometimes the bumps filled with liquid and ruptured, leaving small scars. The 95 percent humidity was perfect for this itchy fungus, which drove him half mad. Dirty water and the constant humidity caused infections inside his mouth as well. Like most of the miners, his breath was rancid. He missed hundreds of little comforts from above, but now his first request was simple: a toothbrush.

Day 18: Above Ground

Pedro Gallo prayed his invention would work. After two weeks of tinkering, Gallo, who owned Bellcom, a one-man telecommunications company, had built a tiny telephone that would fit inside the 3.5-inch limitations of *la paloma* tube. Golborne and other rescue officials had initially ignored the insistent entrepreneur and his "Gallophone," but as one high-tech plan after another failed, Gallo had his opportunity. Minister Golborne was scheduled to speak with the miners, and aides were beginning to imagine the scandal that would erupt if there was not a functioning phone line in place.

After being ignored and having his contraption ridiculed, Gallo was summoned and told to have the phone up and running immediately. He raced to his pickup and pulled out the rustic invention. "They gave me about two hours," said Gallo. The phone was gently packed inside a *paloma* and, along with a half mile of fiber-optic cable donated by a Japanese firm, lowered to the anxious men. Gallo hovered over a cheap yellow plastic telephone set on a flimsy table on the mountainside, with presidential aides and engineers clustered close, waiting as, down below, Ariel Ticona and Carlos Bugueño wired the phone to the Japanese cable. Suddenly, Gallo heard voices from deep inside the mine, echoed and transmitted to the surface above. His invention, which had cost less than $10, was now the key element in the ongoing communications with the miners. Gallo was overjoyed.

Less than an hour later, Minister Golborne arrived and picked up the receiver.

"Hello," said Golborne. "Yes, I hear you!"

A rousing cheer and applause erupted from the rescue workers, who quickly quieted to listen in on speakerphone.

A clear and calm voice was heard: "This is shift foreman Luis Urzúa. . . . We are waiting for the rescue."

"We are starting to drill tunnels and —" Golborne's words were instantly drowned out by a new round of celebration, this time from the trapped miners. The conversation continued with the miners desperately asking about the fate of Raúl "Guatón" ("Fat Man") Villegas, who had been driving up the ramp when the collapse hit. "They are all alive, they made it out," said Golborne, and a chorus of crying and frenetic screams filled the cavern and echoed up to the rescue parties. For weeks, while the world cried for the miners, the miners had suffered for the Fat Man.

The lead psychologist on the rescue mission, Alberto Iturra, listened intently as he stood just behind Golborne throughout the phone call. Dressed in a green reflective vest and safety helmet, his stoic face framed by a trim gray mustache, Iturra neither smiled nor cheered. The medical literature was filled with treatments for claustrophobia and panic attacks and even examples of humans trapped for days in confined spaces. But trapped for months? Iturra knew exactly where to turn. For years he had maintained a network of professional contacts that included a group of esteemed psychologists. Now he would tap that circle. Iturra sent out his own private SOS.

Nursing the miners back from the brink was a delicate task. Starvation had altered the miners' chemical makeup. In addition to burning fat and consuming muscle for energy, the human body, when deprived of food, develops a chemical hierarchy that prioritizes the lungs, heart and brain above now-secondary functions.

Dr. Mañalich, the double-chinned, effusive minister of health, sent a one-page questionnaire to the buried men. Were the miners dying? No. Were they suffering from starvation and loss of body mass? Clearly. How much weight each man had lost was a mystery. Amid the frenzy to deliver a baseline level of comfort, it would be days before a weighing scale could be lowered down to the men, who would then be hung like fish at market, their feet off the ground as they swung from the scale. As the completed questionnaires were returned, the answers revealed fragments of the lost men's experience. Mario Sepúlveda's lost tooth from his climb up the chimney. Victor Segovia's earaches from the blasting piston effect. Mario "Mocho" Gómez revealing he was having trouble breathing, the dust clogging his already sabotaged lungs.

Preexisting conditions, including José Ojeda's diabetes, were now a growing concern. In the absence of ultraviolet light, infections and bacteria could spread through the group in days — if not hours. An emergency vaccination plan was developed to protect the men against diphtheria and pneumonia. An infected tooth could kill. Dr. Mañalich began to research medical history. "We began to look at the old medical textbooks," said Mañalich. "How did doctors treat internal infections like appendicitis before the age of modern surgery?"

Dr. Jorge Díaz, medical director for the Asociación Chilena de Seguridad (ACHS), the insurance company that covered worker accidents at the San José mine, said, "We had hoped they would be alive, but we thought there could be serious injuries and some dead. . . . I knew the miners were tough. So it seemed certain that some had survived." As a specialist in high-altitude injuries and workplace accidents, Díaz was accustomed to logistical challenges. Now Díaz faced the challenge of his career: instead of high altitude, he had to implement a medical protocol for the deeply entombed. Fortunately, Díaz had spent thirty-two years serving miners. He knew the slang, the traditions and the rough world Los 33 inhabited.

The miners were in delicate health. They had lost an average of twenty pounds each, surviving off contaminated water and almost no food. The medical team refused to send the men a solid meal, since a full serving could actually kill them. Known as "re-feeding syndrome," the introduction of a large meal, rich in carbohydrates, to a starved person can induce a chemical chain reaction that drains essential mineral supplies from the heart, leading to cardiac arrest and instant death.

Instead the men were hydrated. Fortt's *paloma* was packed with bottled water and lowered by cable. The first delivery took over an hour. But when the orange PVC tube was hauled up, it was empty — the system worked. *La paloma* was now the life support system for thirty-three men. Anything that was to be delivered had to fit the minuscule dimension of 3.5 inches. Mañalich formed a circle with his hands the size of a lemon and said, "A whole world reduced to this size."

As media reports flooded the airwaves and the Internet with details of the shockingly good news, the world discovered both Chile and the Chilean miners. A new vocabulary was introduced, including the word *paloma* and the phrase "Los 33."

The impression most people had of Chile was either 1970s Pinochet-era human rights abuses or a more modern yet equally superficial association as a producer of tasty — and inexpensive — wine.

Now, the eyes of the world shifted to this previously obscure corner of northern Chile. Plane flights and hotel rooms sold out. The rental price for a motor home — a favorite on-site sleeping quarters for foreign TV crews — soared 300 percent. English translators throughout the region were booked solid. Hundreds of reporters rushed to the scene for a rare moment when the lens of the world focused on a story with neither blood nor violence.

The Chileans had first found the men at a depth twice the height of the Eiffel Tower. Now they had a second Mission Impossible: to keep the men alive for another four months, until Christmas, the time it was expected to take to dig the men out.

At his office in Berlin, Pennsylvania, Brandon Fisher watched the TV screen in amazement. The bearded thirty-eight-year-old couldn't believe what he was hearing: three to four months? As president of Center Rock Inc., Fisher oversees the design, manufacture and delivery of drill systems that cost up to $1 million. Fisher didn't think it was necessary to drill through the rock. His company specialized in manufacturing pneumatic hammers that smash rock twenty times a second, effectively pounding the rock to pieces.

In 2002, Fisher participated in a rescue at the Quecreek mine, a Pennsylvania coal mine where nine miners were trapped for seventy-eight hours when 50 million gallons of water flooded in. The rising water threatened to drown the trapped men. Fisher participated in the drilling operation that saved the miners as water lapped ever higher in the flooded tunnels. Now he flashed back to the Quecreek operation. Mine collapse. Trapped men. Emergency drill operation. Fisher instantly saw a role for Center Rock. Fisher wanted to volunteer. He began looking for flights to Chile.

Late that same afternoon, a millionaire drove into Camp Hope at the wheel of his gleaming yellow Hummer. With his tailored Ermenegildo Zegna suit, cuff links and rolls of bleached-blond curls bouncing to his shoulders, Leonardo Farkas was unmistakable. To Chileans, the forty-three-year-old mine owner was an exemplary businessman; he'd never have let such an accident occur in one of his mines. Farkas's mining companies Santa Fe and Santa Barbara are open-pit iron mines widely recognized as operations that prioritize worker safety, fair wages and profit-sharing plans. A job with Farkas was a guarantee of top living and retirement benefits. "You have to wait for someone to die to work there," joked Mauricio, a taxi driver in Copiapó who applied in vain for one of the two thousand employment slots with Farkas's mining operation. "It is like a big family; everyone wants to work there."

Farkas is a legend in Chile for his spontaneous charity, ranging from million-dollar donations to the Teletón, a Chilean fundraiser for disabled people, to the afternoon he walked by a swimming pool filled with university students and offered a reward to the fastest swimmer in the pool. The first one across the pool would receive a check for one million Chilean pesos — the equivalent of $2,000. Sports are an important part of education, said Farkas, who minutes later wrote a check payable to Eduardo Hales, the astonished winner. Restaurant waiters who served Farkas were often rewarded with tips in the thousands of dollars.

Stepping from his Hummer, flashing his curls and gleaming white teeth, Farkas looked like a lounge singer from Las Vegas teleported to the wrong desert. Farkas began to hand out plain white envelopes, one to each family. Inside was a check for 5 million Chilean pesos — roughly $10,000.

"From the first day, my company has cooperated here," said Farkas in a brief statement in which he hinted at but did not explicitly mention the boxes of sandwiches his company had regularly delivered to the rescue team. "We bought parkas and hats for this cold weather. Not all our contributions are public or told to the press." Farkas then announced a campaign to raise $1 million for each miner — a call for businessmen and citizens alike to "reach into their pocket" to assure that the men would never again need to work. "I don't want that when these guys get out . . . they have economic worries," said Farkas. "I am not here to offer them work; I am here to offer them something better than work — that every family has a million dollars." Grateful family members promised to deposit the checks and noted that Farkas had wisely made the checks payable directly to the individual miners, avoiding ugly disputes.

While Fisher and Farkas organized their

separate plans to help the trapped miners, Alejandro Bohn, co-owner of the San José mine, incited a new storm of criticism when he gave an interview on August 23 to the Chilean radio station Cooperativa and announced that the company "is tranquil" about the possible legal fallout from the mining accident.

"We never had any forewarning of this kind of catastrophe. The workers were trained and had the security equipment so that they could deal with this kind of event and they would have the necessary protection," said Bohn, who hinted that the company might stop paying salaries for the thirty-three trapped workers and another three hundred company employees. "We have spoken to the authorities with respect to searching for solutions to continue operating. Unfortunately, for the moment, they — like us — are focused on the rescue of our workers."

Asked if he planned to apologize in any way to the miners and their families, Bohn hesitated. "It is necessary to be cautious. The investigation must be advanced to see if anything could have been done beforehand." The mine owner also refused to testify at an upcoming hearing before an investigative committee of the Chilean congress.

Minutes later Minister Golborne led a cavalry-sized attack against Bohn. "I find these statements incredible. I heard them and was really surprised." Golborne then blamed the owners of the San José mine for failing to install a safety ladder in the ventilation shaft. "We could have avoided this whole drama," said Golborne, adding that the accident highlighted "a very important lack of attention to security" inside the mine.

Senator Alberto Espina also lashed out at Bohn and accused San Esteban Primera S.A. of "bad management, not fulfilling labor laws, provoking a dramatic situation and, finally, distancing itself and saying we don't have money to pay the salaries. It is quite incredible."

"In the very least they could come before the investigative committee and explain what happened," said Frank Sauerbaum, a congressional representative who said the mine owners have "systematically refused to assume their responsibility." Sauerbaum also noted that the miners were alive "thanks to the professional and steady work of the government. If the company that owns the mine had been in charge of the rescue operation, this story would have been completely different."

Day 20: Wednesday, August 25

Luis Urzúa was now busier than he had been in weeks. The authorities above were directing all their messages to the shift foreman, a clear strategy to reinforce Urzúa's much debilitated power. President Piñera called Urzúa to hear firsthand how the men had survived. "How we tried to escape this hell . . . That was a terrifying day," said Urzúa as he described to Piñera how the men fought to escape the initial cave-in. "It felt like the whole mountain fell atop us and we did not know what happened." Urzúa then pleaded with Pinera: "The thirty-three miners who are here inside the mine, under a sea of rocks, are waiting for all of Chile to get us out of this hell."

Urzúa agreed to make a video for the government. A camera would be sent down and the men were to film their living conditions and conduct a brief tour of their remarkable world. As the conversation continued, the miners relaxed and the dialogue became more informal. They asked the president to send a special treat for the upcoming bicentennial celebration on September 18: "a glass of wine."

Day 21: Thursday, August 26

As the miners prepared to sleep, a nine-minute video was released by the Chilean government, broadcast at prime time on a Friday night in Chile. A window was opening into their underground civilization. It was the miners' first TV appearance. As the news video zipped around the world, the response was incredible. The world was stunned.

Florencio Ávalos held the camera while Sepúlveda slowly panned inside the tiny cave that was the safety shelter. The crude, irregular rock walls. The rusted oxygen tank. The cracked tub that served as a holding bin for a jug of water. The tattered medicine chest that was no bigger than a knapsack and the medications that had long since passed their shelf life.

Huddled like frightened animals, few of the men looked at the camera. Sepúlveda tried to cheer them up, to stir their group spirit. Few of the miners responded. Pablo Rojas tried to speak but choked up. Other men lay prone on the floor, avoiding the camera. Exhaustion hung heavy in the crowded shelter; tired eyes stared off into nowhere. They looked like antique black-and-white pictures of traumatized soldiers.

Dirt and overgrown facial hair disguised the men in a cloak of universal suffering. Claudio Yañez looked barely strong enough to stand, his ribs rippling out of his chest. Like a platoon of weary guerrilla fighters, the men exuded an aura of heavy trauma. Death, or the sensation that death was near, gave the video a haunting humanity.

Some men wore orange mining helmets, but few wore shirts. Sweat rolled down their bodies in rivulets. Packed into the 160-square-foot safety shelter, the miners looked distraught. Sepúlveda continued with his cheerleading performance, joking that a miner had found a new box spring and mattress and cajoling the men to share a few words with beloved family members. Zamora rallied his energy to thank the families. "We know how you fought for us." Zamora paused to dry his tears. "And we all applaud you." The cheers were brief.

At the end of the video, the miners began to sing the Chilean national anthem, their voices ringing out despite their obvious exhaustion. Whatever else the world would take away from the first sight of the miners, few would doubt that they were united.

The video was a virtual tour of the miners' secret world. While many of the miners were shown lying down and appeared shy in front of the camera, Sepúlveda, with humor and eloquence and brimming with confidence, put on the performance of a lifetime. He prodded the men one by one to address their families, to send a few brief words of hope and greetings. The video was a shockingly positive summation of the miners' fragile existence and a proud declaration of survival.

Sepúlveda's role was not a stroke of luck but a media-savvy strategy: the Piñera government had worked with the miners to appoint Sepúlveda as host. "We had to ask the miners not to put Florencio Ávalos on TV but to use 'the artist' [Sepúlveda]," explained Dr. Mañalich, the health minister. "It was a very difficult negotiation." The Piñera government wanted to showcase the miners to the world as heroes, human trophies highlighting the president's inspiring and entrepreneurial spirit. But this media strategy required select and careful editing. The video was carefully censored; images of the men's fungal infections were edited out. Scenes of sobbing miners were never shown.

Day 22: Friday, August 27

A flood of letters came up from below, handwritten notes detailing the men's unique world. Psychologists and family members could now begin to piece together routines and rules for this miniature society. The miners detailed the logistics of their three working groups of eleven men, revealing how each group took turns for an eight-hour shift, in an ongoing fight to survive underground. "We have three groups, Refugio [Refuge], Rampa [the Ramp] and *105* [meters above sea level]," wrote Omar Reygadas in a letter to his family. "I am head of one [Refugio]." Each group had a leader, a "*capataz,*" who reported directly to Urzúa.

As the men began to recover their strength, a daily schedule was organized. Rescue leaders feared that with food and water no longer scarce, without strict schedules imposed from above, the men would relax all day long and social cohesion would disintegrate in a textbook example of "idle hands are the devil's workshop." Led by the *capataz,* each group was given daily tasks. For the morning shift, the day began with a 7:30 AM wake-up, breakfast at 8:30 and a morning of chores, some sent from mining engineers above, others simple obedience to common sense.

To the surprise of professionals both in Chile and at NASA, the miners had developed a protocol of routines and tasks that turned the seventeen-day experience into an extension of their everyday routines. Instead of abandoning their individual roles, many of the men adapted and employed their mechanical and electrical skills to construct new inventions that were key to their survival. The continuation of routines had allowed the men to avoid a sensation of helplessness. "Our goal is to help them help themselves, not to treat them as sick," said Dr. Llarena.

With their energy rebounding, the miners began to reinforce weak walls, clear debris and divert the streams of water seeping into their sleeping areas. The paloma tubes connecting the men to the surface were lubricated with water, creating a stream of muddy gunk that constantly dripped into their world. Letters from the men were stained with drops of sweat and blotches of brown mud — permanent reminders of the 90 percent humidity and 92-degree air temperature inside the mine. But now they were receiving shampoo, soap, toothpaste and towels — a five-star upgrade in comparison to just days earlier.

The men organized security patrols along the perimeter of their sleeping and living quarters, a constant vigil for signs that the notoriously unstable San José mine might again be giving way and trapping them in an even more confined space. The miners feared a small stream of rocks could give way, then expand avalanche-like into a full-scale collapse. The men spent hours every day "*acunando*" — using long-handled picks to clear large rocks from the roof of the mine.

"They will hide like rats and seek shelter at the first major movement of rocks," said Alejandro Pino, a lead organizer of the rescue operation with the Asociación Chilena de Seguridad (ACHS). "These are experienced miners. At the first sign of major movement, they know where to hide."

With paloma deliveries arriving every forty minutes, the palomas created a constant chore for the trapped men. Six miners were assigned as *palomeros,* a new Chilean word meaning "pigeon catchers." The *palomeros* were tasked with receiving the 10-foot metal tube, unscrewing the cap, pouring or shaking out the contents, and stuffing in the latest letters and messages, then waiting for the torpedo-like tube to rise out of sight.

"We only give them a short time; they have to complete the paloma operation in ninety seconds," said Dr. Mañalich, the minister of health. "It could be there for ten minutes, but we give them less than two minutes so they have to complete routines. . . . Yesterday they told us, 'We have never worked this hard in our life.' That is a very good sign. They should not stop at any moment. They have to be working for at least eight hours during the day."

Even when it was not their turn, the miners began to wait at the paloma station, either to receive a cherished letter or out of sheer curiosity about the gadgets, goods and never-ending barrage of incoming packages. Thanks to the increasingly efficient delivery system, four days after contact had been made, the miners had a projector, new head lamps and a stash of fresh water in their refuge. Rescue workers urged the men to stockpile fourteen days' worth of food. "They are starting to have a strategic reserve," said Pino of the ACHS.

Food deliveries and meals took up a chunk of the day. Lunch delivery started at noon, and it took a full hour and a half for all the meals to arrive. "When they finish lunch, they have a general meeting, and in this meeting they start their prayers," said Dr. Díaz.

José Henríquez, as usual, led the daily prayer. "Don José" lived for Jesus and his daily sermons. What began as a small prayer group had by now turned into a full-fledged evangelical conversion. Twenty men regularly went to his mass, sometimes more. Henríquez could now count on Florencio Ávalos, the group's official cameraman, to record his sermons.

Pedro Cortés and Carlos Bugueño were

appointed as sound technicians and put in charge of maintaining the phone lines for conference calls scheduled for the early afternoon.

Nineteen-year-old Jimmy Sánchez, the youngest of the group, became the "environmental assistant" and, together with Samuel Ávalos, roamed the caverns with a handheld computerized device to measure oxygen, carbon dioxide levels and the air temperature. Every day Sánchez and Ávalos took the readings off the Dräger X-am 5000 and sent reports to the medical team above ground.

With basic needs including food and sleeping quarters now organized, the men began to fill bureaucratic and cultural positions. José Ojeda, now known worldwide as the author of the famous first note, was named the official secretary. Victor Segovia continued as the group's official chronicler, penning daily accounts in an ongoing log of the men's predicament.

Within days of the initial contact, rescue officials appointed Yonni Barrios as the group's doctor, recognizing a position Barrios had already assumed for himself during the first seventeen days. He quickly recruited Daniel Herrera, who was given the title "assistant paramedic."

Of all the men tasked with keeping the group functioning, Barrios was perhaps the most crucial. He vaccinated the entire group against diphtheria, tetanus and pneumonia, and with fungal infections and bad teeth at the forefront of medical problems facing the miners, Barrios found himself at the center of an unprecedented experiment in telemedicine.

Apart from daily medical rounds, Barrios had an hour-long conference call every afternoon in which he received messages from the medical team.

"Yonni, can you hear me?" yelled Dr. Mañalich during a medical conference call conducted by a telephone hooked up to a 2,300-foot cable. "Yonni, have you ever pulled out a tooth?"

From far below, the crackle of Barrios's voice arrived topside. "Yeah . . . one of my own."

The doctors looked at each other in surprise, shocked by the miner's humble reality. "If we have to ask you to pull a tooth and send you sterilized equipment, could you?" asked Mañalich, who promised to send video instructions on how best to extract an infected molar. Mañalich sent a friendly warning to Barrios: "Remember, Yonni: tell the men if they don't keep brushing their teeth that you will soon be ripping their teeth out."

Barrios had one other important task. "We needed him to measure the men. We needed their circumference in order to find out if they would fit through the small rescue hole now being drilled," said Dr. Devis Castro, a surgeon with advanced studies in nutrition.

Above ground Barrios had an even more complicated operation — keeping separation between his lover and his wife, both of whom were battling for him in public attacks that had the media in a frenzy. Below ground, the men never ceased to rib Barrios about the controversy. In the cloistered world of miners, jokes and humor continue nonstop. Nothing is sacred. Instead of respecting Barrios's delicate dilemma, the miners plumbed it for every ounce of laughter, teasing and taunting without malice — simply as part of the daily conversation.

Day 24: Sunday, August 29

Six days after first contact via Pedro Gallo's rudimentary phone, now the main channel of communications with the miners, demands from below increased. The miners wanted, needed, pleaded to speak to their families. The rescue leaders scheduled very brief voice contact: each family would have sixty seconds with their loved one, as recommended by the psychologist Iturra.

The miners were indignant. After having spoken with President Piñera and Minister Golborne for well over an hour in total, now they would collectively receive just thirty-three minutes for what was their most important call to date? When the calls began, so, too, did a new round of problems.

"I was talking on the phone and Iturra was saying, 'Cut, cut, cut' and I was like, what are you talking about? That is not even one minute. Then he said, 'Cut or I cut you off.' I thought, what an asshole; that gave me an idea of his mentality." Samuel Ávalos accused Iturra of being overly strict and possessive of the miners. "He wanted to impose his terms on the group. We were never going to accept that. . . . We were a group, for better or worse a family."

Initially the miners agreed to a two-hour daily conference call in which Iturra and doctors peppered them with questions — an attempt to build a psychological profile of the group and its individual members. As the miners regained weight and strength, however, their antagonism to the daily sessions increased. "They say they are not sick and they do not want to talk to doctors or psychologists," said Dr. Díaz.

The new level of communications also began to seed a crop of controversies and conflicts. Family feuds above ground threatened to spill into the letters and conversations with the miners. No one knew how much more mental stress the miners could take — one miner losing his mind had the potential to infect the entire group. Rescuers worried that panic attacks or violence could engulf the miners in a collective state where reason and order vanished.

With dozens of letters flowing in both directions every day, the psychological team led by Iturra instituted a strict policy. All letters from the miners would be read before being released to the families. Similarly, any letter intended for the miners would also be read by a team of psychologists who spent the days going through stacks of tightly folded, handwritten missives.

Nick Kanas, a longtime adviser to NASA, was critical of the censorship and Big Brother mentality. "I would not screen anything . . . otherwise you are setting up a basis for mistrust. The miners will then start asking, 'What else are they hiding from us?' They will know they are not getting the full story and will want to know why."

As it was, tensions rose quickly. José Ojeda did not believe that letters were lost or delayed, as government officials tried to explain. "This is like a jail; they censor everything," he wrote. "We were better off before we had communications." That letter was never shown to his family but was stashed away by the psychologists.

"Sometimes they would add words or they would rewrite the letters," said the miner Carlos Barrios. "I know my grandmother's handwriting." Barrios began to talk about a strike. The miners would present a united front to the invisible commanders above. For Barrios, the entire incident highlighted psychologist Iturra's patronizing attitude, an attitude that united the men. "They thought we were ignorant," said Barrios. "They never understood us."

EIGHT
THE MARATHON

Day 26: Tuesday, August 31

As the gray van threaded its way through the crowd of cameramen and photographers at Camp Hope, family members of the trapped miners lined the roadway and cheered. Inside the van, six specialists from the National Aeronautics and Space Administration (NASA) stared out in wonder. Having been trained in the comparatively sterile and highly regimented bureaucracy of the U.S. space program, the sight of dozens of women shouting at them in Spanish while hundreds of journalists jostled to take their picture was like arriving on another planet.

The news of the miners' survival underground for seventeen days had stunned the world, as had the Chilean expertise in drilling holes and marshaling mining equipment that had led to contact being made with the trapped men. But now with the men beginning to receive meals and medicine, an entirely new challenge arose: maintaining their psychological health. Rescue leaders at all levels were floundering in uncharted regions of the human psyche. In acknowledgment of the unique characteristics of the San José mine disaster, President Piñera sent aides to find expert consultants with relevant experience. They came back to the president with two recommendations: astronauts and submariners.

Chile's space program was limited to one man, Klaus von Storch of the Chilean Air Force. Von Storch was a die-hard optimist who had patiently sat on NASA's standby astronaut list for more than a decade before giving up. Although the Atacama Desert placed Chile at the forefront of world astronomy, manned space flight was light-years away from the nation's fiscal reality. So with no local data to call upon, the Chilean embassy in Washington, D.C., contacted officials at NASA, who were delighted to share decades of studying human behavior in confined, stressful situations. The team at Camp Hope included Dr. Al Holland, a psychologist with vast experience in extreme living conditions ranging from the deep space of Apollo missions to deep-freeze environments in Antarctica.

The NASA specialists huddled with Chile's

recently formed team, which included psychologists, nutritionists, mining engineers and Renato Navarro, a commander with the Chilean submarine fleet who had been brought in to share his experience of managing men in confined environments. "The submarine has water outside; the miners have a seven-hundred-meter [2,300-foot] high column of rock," he said. "The sense of confinement is the same."

Known to psychologists as "Situations of Extreme Confinement," the living conditions of the thirty-three miners presented so many logistical and mental health issues that the support staff at the mine now swelled to include a total of three hundred professionals, including a physics professor, a mapmaker and an avalanche survivor. Also on staff was Edmundo Ramírez, a chef brought in to prepare the meals sent down to the miners. The visiting dignitaries from NASA were the latest in a stream of foreign experts, but even with ten professionals for each trapped miner, many questions could not be answered.

"This is an unprecedented situation and effort," said Michael Duncan, a NASA psychologist, speaking inside a tent at the San José mine. "To my knowledge, never before have this many men been found so deep underground. The fact that they were found such a long time after the collapse and found alive was remarkable."

The NASA officials lauded the Chilean rescue effort and suggested minor changes to the protocol, including additional vitamin D and better artificial lighting to stimulate the body's reaction to the cycles of day and night. The NASA team also emphasized that simple daily activities like playing cards, reading and watching movies were crucial to avoiding a monotonous existence. NASA officials refused to release many details of their final five-hour briefing, but participants in that meeting with NASA said the U.S. space agency had vigorously promoted the importance of organizing the miners in a strict — almost corporate — hierarchy. Voting and group decision making had worked fine for seventeen days, but now, NASA stressed, the men needed to be prepared for a race with different stages — in the words of NASA, "a marathon."

NASA officials also told the rescue leaders to prepare for a rebellion. "They said that during one of the Skylab missions, the astronauts had an argument with their commanders [and] became so upset that they cut off communications with the commanders," recounted Dr. Jorge Diaz. "For a day the astronauts orbited [Earth] and no one could contact them."

The Chilean psychiatrist Dr. Figueroa echoed this sentiment. "Following the euphoria of being discovered, the normal psychological reaction would be for the men to collapse in a combination of fatigue and stress," he explained. Dr. Figueroa had been hired by the Chilean Ministry of the Interior to report on the mental health care being provided to the miners and their families. "There are approximately fifteen percent of the miners who could develop long-term psychological damage from this event. This is where the government is very dedicated to strongly supporting the people to prevent these long-term psychological problems. The most important thing is to open a channel of communications, a prescribed time when the miners can send messages."

Letters had already proven to be a huge psychological boost to both the families and the miners. Among the first requests from the trapped men were pens and paper. The Chileans had also phased in a phone system with the miners; this was to be followed by a video conference system. But open communications also meant a loss of control. What if a wife decided to ask her miner husband for divorce? Was this really the time to fight over household bills and finances?

Day 27: Wednesday, September 1

From a distance the rescue site at the San José copper mine looked like a construction site gone mad. Huge cranes rattled twenty-four hours a day, transporting metal tubes the length of a ship's mast with ease. Cement trucks, bulldozers, backhoes and robotic mine machines that looked like insects prowled the mountainside. Parking lots were filled with supplies ranging from a field of drill bits to twenty-eight pallets packed with charcoal. Placed in an oil drum and lit, the burning charcoal served as a night-light and heater for the estimated twenty policemen stationed as sentries along the hillside.

Shifts of helmeted men, their huge hands grubby and their faces slow to smile, were evidence of the arduous task that had gathered hundreds of rescue workers for the past four weeks. Inside the mess tent were men from Brazil, South Africa, the United States and Canada who joined hundreds of highly trained Chileans. These rescue workers had missed their children's birthdays and abandoned their families to fly to the Atacama Desert to help. They volunteered for twelve-hour shifts to try to save men they didn't know, men they might never meet.

Caravans of 4x4 pickups arrived with food, machinery and donations. "We are here to provide support to the families and the kids. Every four or five days we bring milk and yogurt to these one hundred and eighty people," said Adolfo Duran, distribution supervisor for Soprole foods, pointing to stacks of yogurt cartons and crates of milk. "The feeling of fraternity has been augmented heavily this year; first we had the earthquake and now this. Personally, I feel like our nation has become much stronger this year."

Down the mountain, below the police checkpoints, family feuds erupted and became part of the media circus. Hundreds of reporters trapped behind security lines had little to do but interview one another or speculate. How many of the married miners had lovers? Were the trapped men having sex? Was the operation really going as smoothly as the Piñera government was portraying it?

Despite the outpouring of support and help, Camp Hope was no superficial love fest. Family feuds erupted and tears flowed in disputes. "Yonni doesn't want to come out of the mine," joked a doctor working at Camp Hope as he described sorting out the ongoing love triangle that continued to ensnare miner Yonni Barrios in a secondary net of entrapment. His longtime wife and his longtime lover continued to battle — even destroying the photos placed at his official shrine. In family after family, the story was the same: long-lost daughters and sons flocked to see the father who had never been a father, a painful and touching demonstration that the heart tugs at even long-frayed blood ties.

Local government officials realized that Camp Hope would continue to grow. The population was now five hundred and new "neighborhoods" sprouted weekly as journalism teams arrived to stake a claim to a piece of turf and a chance to find nuggets in a story the entire world was now watching. In 2000, when the *Kursk,* a Russian submarine with a crew of 118, sank to the bottom of the ocean, the world's media was fixated on the plight of the trapped sailors, who slowly died, their story measured by the ever fainter *"tap — tap tap,"* a Morse code message played out on the submarine's shell. A decade later, almost to the day, the Chilean mining drama became arguably the world's biggest ever multimedia tragedy. With the completion of a fiber-optic connection to the men, digital video cameras that were sent down *la paloma* as well as entertainment systems, including a video projector and MP3 players, allowed the thirty-three miners to quickly become among the most wired and media-savvy disaster victims in human history. Two months after the collapse, the number of hits on Google for "Chilean" and "miners" hit 21 million.

The drama of the Chilean miners was fast becoming a daily staple in the world's entertainment diet.

Camp Hope now had zones for children, community bulletin boards, and scheduled bus shuttle services to nearby cities as well as an evangelical preacher's stage with amplified prayer and scratchy speakers set up just 10 feet from the international press tent. While reporters and producers filed news reports, they were often serenaded by cries of faith, promises of salvation, and reminders not to forget the "thirty-fourth miner," Jesus Christ.

While Chilean officials continued to caution that huge technical and logistical challenges lay ahead in removing the men from the mine, families laughed and prepared barbecues, at peace with the knowledge that the miners were alive. With bonfires and an abundance of positive energy, the camp felt less like a refugee camp and more like a scaled-down Chilean music festival. Live performances abounded. At the keyboards the famed Chilean pianist Roberto Bravo, surrounded by a ring of family, uncorked what he described as the performance of his lifetime.

"I can breathe easy now. There's no more doubt," said Pedro Segovia, thirty-eight, brother of the miner Darío Segovia. "Before, we didn't know if the machinery could really find them at seven hundred meters [2,300 feet]." As he sucked on a lemon, dousing it regularly with salt, Segovia described the San José mine as a death trap. "I worked there for a year. It was always a dangerous place to work. All of us who went in there would wonder, Will we make it out? Once a piece of the roof, a one-hundred-kilo [220-pound] rock, fell on me. Luckily it shattered on a protective screen and only bruised my back."

Pedro Segovia took shifts with family members and friends to stand watch in their family tent, where a solitary candle burned amid images of Jesus and the Virgin Mary. The family's vigilance was not for fear of robbery. Camp Hope was the kind of place where lost cell phones were cordially returned to grateful owners. The Segovia family kept one member awake out of respect for Darío. He was directly below them, trapped. How could they all be asleep?

Adjacent to the Segovia tent, a group of children played with the candles in the shrine to their grandfather, Mario Gómez. With pencils and crayons, they drew simple pictures of cars and solemnly stacked the drawings next to his photo before running to play in the rock piles that dot the otherwise barren hillside.

Camp Hope was now becoming a community. Although each family set up its individual home and daily routines, a common cause and purpose had given the crowded living conditions an air of civility. Among the family members there were few secrets. The combination of abundant free time and a common passion meant that news traveled briskly in the small camp.

Carolina Narváez, wife of Raúl Bustos, was becoming familiar with tragedy. Six months earlier, trapped at the epicenter of the 8.8 magnitude earthquake, Narváez and Bustos watched a tsunami destroy the shipyard where he worked. Working at the San José mine was always intended to be a temporary stint until Talcahuano, Bustos's hometown, 745 miles south, was rebuilt. "Nobody has ever lived this long underground. I can't be weaker than him," said Narváez, sitting on a rock and smoking a cigarette. Behind her a poster showed Raúl, staring out, his face grim. Narváez held no illusions that they would survive the ordeal unscathed. "I know the Raúl who went in there is not the Raúl who will come out."

In a nearby campsite, just 65 feet away, Nelly Bugueño was practically celebrating that her son Victor Zamora had been trapped. Always critical of her son for rushing and suffering day-to-day stress, Bugueño said the entrapment had forced Victor to look inward. She read and reread his letters with wonder. As a lifelong miner, Victor had never shown such a talent for bold, emotional writing. This was definitely not the same Victor she had first raised and then watched develop into a lifelong miner.

"He found his second self down there. He has discovered that he is a poet. Where did all these beautiful sentiments come from? Did they sprout?" Bugueño smiled, her petite stature overshadowed by her immense pride. "I don't want him working in mountains anymore. He should write songs, write poems."

In a nation that produced the Nobel Prize–winning poets Gabriela Mistral and Pablo

Neruda, it is not surprising that the men named Zamora the miners' official poet. Zamora's rhyming compositions were often one-page homilies to the rescue workers. His combination of hope, gratitude and humor quickly made them among the most-read messages from below. Even after multiple readings, Zamora's poems brought tears to the eyes of Pedro Campusano, a paramedic working at the paloma station. "When the first one came up, I read it and got halfway through; I couldn't —" Campusano's eyes filled with tears. "When I read it . . . it fills me with emotions."

Despite the initial euphoria over finding the miners alive, extricating them from the mine — what Chilean engineers dubbed "the Final Assault" — was a daunting challenge. It would entail a three- or four-month effort to drill a hole 2,300 feet to the trapped men, and design a system to haul them one by one up from the refuge. In recognition of the unprecedented nature of the challenge, the Piñera government opted for multiple rescue strategies, with deliberately varied technologies. The two exhaustively complicated drilling plans had deceptively simple names: Plan A and Plan B.

Plan A was designed around one of the

world's largest drills, a sophisticated Australian rig known as the Strata 950 raise borer. The raise borer was capable of drilling a 26-inch-diameter hole as deep as 2 miles, at a cost of $3,000 to $5,000 for every meter (3.28 feet) drilled. Only six such machines existed; fortuitously, one had been located in Chile. To rescue the trapped miners, the engineering plan called for the raise borer to drill straight down to the men. First the machine would carve an 18-inch hole, and then a second, wider bit would enlarge the tube so that the men could be hauled out in a rescue capsule. The drilling was slow but sure; in four months — by Christmastime — the tunnel would be finished. Experts all agreed that the Strata 950 could finish the job. But after such a prolonged confinement, would the miners be sane or even alive?

Chilean authorities were now flooded with hundreds of proposals to save the miners. With barely a moment to pause, the authorities jumped aboard the strategy used to save the miners in the Quecreek mine in Pennsylvania. The plan called for using one of the original boreholes and widening it using a powerful American-made drill known as the Schramm T-130. Dubbed Plan B, this scheme offered the possibility of rescuing the men in less than two months. However, there was no guarantee that the techniques that worked at 230 feet could now be extended to save miners trapped at ten times that depth.

Day 29: Friday, September 3

Brandon Fisher arrived at Camp Hope with a singular mission: to help guide Plan B. The tireless engineer was now reunited with members of the same team that eight years earlier had saved the miners in rural Pennsylvania. Could he repeat the miracle?

James Stefanic, president of Chilean operations for the U.S.-Chilean company Geotec Boyles Brothers, located the same model drill as the one that had been used in Quecreek — the Schramm T-130 — at the Doña Inés de Collahuasi mine in northern Chile. The 100,000-pound rig is highly portable and comes with five axles, meaning it was easy to transport and could be set up almost instantaneously. Arrangements were made to bring the rig to the San José mine.

Plan B might also have stood for "Blind." There was no way to guide this drill. Fisher was the key. With his Center Rock factory on call in Berlin, Pennsylvania, Fisher and his eighty-person company would craft a solution. Fisher was sure his team could design and manufacture a drill with a small snout on the tip that would fit snugly into the borehole, essentially keeping the now bigger drill on course.

Still, on many fronts Plan B was experimental — for one, the drill had never been used for a rescue so deep. "One of the most important things when you drill is to know exactly what the drill is going to weigh," said Mijail Proestakis, an engineer on Plan B. "It is easy to go down, but you have to remember to be able to pull up everything." Engineers were cautiously optimistic that the machine could handle the weight of the entire drilling shaft — an estimated 48 tons.

The Chilean Embassy in Washington, D.C., convinced United Parcel Service, the giant shipping company based in Sandy Springs, Georgia, to coordinate a massive rush shipment. Twenty-seven thousand pounds of drilling equipment were flown from the Pennsylvania iron belt to the remote Atacama Desert. The UPS Foundation, a philanthropic division of the $50-billion-a-year shipping behemoth, picked up the tab.

A key part of Plan B was still missing: the driller. Despite technical advances in drilling systems and GPS technology, the Schramm T-130 still needed a captain to guide the mission. Stefanic knew exactly who he wanted at the helm.

Jeff Hart, a towering forty-year-old sun-blasted oil worker from Denver, Colorado, was an expert at finding buried treasures. Hart was regularly flown to ugly corners of the planet to guide drills.

At the time, Hart was drilling for the U.S. Army in Afghanistan. In that land so studded with minerals, oil and gas, Hart had been hired to find the most valuable underground lode of all: fresh water, the new Afghani Gold.

The initial message to Hart was stark. A mine had collapsed in South America. Thirty-three miners were alive but buried 2,300 feet deep — at the bottom of a gold and copper mine. Was he willing to come and try to drill them to safety? Hart agreed and, like a character in a James Bond film, was "extracted" from deep inside rural Afghanistan and flown to Dubai, then Amsterdam and on to Chile. Asked why he chose Hart, Stefanic was clear: "He is simply the best."

With Hart set to take the controls of Plan B, the competition between the two teams escalated. Engineers on-site began placing bets on which rescue operation would reach the miners first. Glen Fallon, a towering Canadian who was the lead operator on Plan A, said he welcomed the competition. "There was a global SOS that went out on this. Now I get emails every day from people who want to volunteer, fly to Chile to help out," he said. "Even my competitors are offering to help. In this race, there is only one team."

Day 35: Thursday, September 9

Jeff Hart felt at home with the controls of the T-130 Schramm drill, a machine he had operated for thousands of hours. Using levers and foot pedals, Hart worked standing up, rarely removing his dark sunglasses, his ears wrapped in bulbous yellow protective gear. A rag hung from the back of his helmet, shielding his neck from the Atacama sun. After traveling halfway around the world, Hart was now on track to find his most valued target ever: a group of treasured humans. For days he rarely moved from his workstation at the Plan B drill. He drilled for ten hours a day, the passage of time measured by the growing collection of oil and mud stains that covered his jumpsuit. Then, on September 9, just the fifth day of operation, Plan B stalled.

Meanwhile Plan A continued to grind slowly into the mountainside. The huge machine spun and smashed through 490 feet of rock. While Plan B drilled far faster, it had to first make a small hole, then drill again to widen it enough for a human to squeeze himself out of the hole. Plan A was the tortoise — slow but steady — as it continued drilling a shaft easily wide enough to rescue the men.

Hart watched in confusion as the air pressure collapsed and the drill spun but no longer cut into the rock.

At 879 feet, the operation was stalled. Hart tried to decipher the signals from below. The engineers had no choice but to stop the drilling and remove the drill shaft, segment by segment, until they could inspect the hammer. The evidence was obvious: the drill head was shredded. Football-sized chunks had been torn off the tungsten-steel shaft. A video camera that was lowered down the hole revealed the missing pieces had become entangled with iron. Faulty maps had led the engineers to design a drilling route that passed through a layer of rods used to reinforce the mine. Now those rods had sabotaged the rescue tunnel.

Day 36: Friday, September 10

The engineers lowered huge magnets down the drill hole to try to remove the metal chunks, but that effort failed. An attempt to batter and loosen the trapped shards was also unsuccessful. The metal was stuck. The rock held the hammer fragments tight, like a fishing lure snared at the bottom of a lake.

Igor Proestakis, a twenty-four-year-old Chilean engineer, had been brought to the rescue site by his uncle Mijail, one of the lead engineers for the entire rescue. Among the youngest of the engineers on-site, Proestakis heard about the problem of the trapped chunks of the hammer and began to design and draw a solution. Proestakis remembered from his university classes the decades-old technique of recovering material lost in the depths of a mine: lowering an open metal jaw with sharp teeth to the bottom of the shaft and placing it around the target — in this case the tungsten pieces. Extreme pressure was then applied atop the metal jaw, like a giant's foot crushing an aluminum can. The pressure from above forces the sharpened teeth to slowly bend shut, trapping the "prey." Known as "la Araña" ("the Spider"), the technique was crude but time-tested. Still, Igor's repeated suggestions to use the Spider were ignored.

With Plan B stalled, rescue leaders panicked further when Plan A was forced to stop drilling. A leaky hydraulic hose required urgent attention.

With both drilling machines now stalled, the miners were serenaded by the most fearful sound in the mine: silence. Not a single machine was heading toward them.

Day 37: Saturday, September 11

With Plan A advancing slower than expected and Plan B stuck — perhaps fatally — a wave of gloom and fear swarmed over the camp. Were the miners cursed? Was all this rescue effort nothing more than a prelude to the inevitable death of yet another group of mine workers? The government was determined to keep the rescue effort moving forward and had already invited a third rescue team to the San José mine — Plan C.

The arrival of Plan C, a massive petroleum drilling operation, kicked off a brief flurry of cheers and flag waving at Camp Hope. The press corps — now chafing at their inability to watch the rescue effort firsthand — rushed to film the convoy of forty-two trucks inching up the gravel roads, loaded with tubes, towers, generators and so much machinery that the platform to support it was 328 feet long, the size of a soccer field.

The rig was donated by Precision Drilling, a Canadian company that specializes in deep perforations to find pools of oil. The rig had been stored for two years in a warehouse in Iquique, a northern port city a thousand miles north of Copiapó.

With scrap copper now reaching a record price of $6 a kilo, copper crime was on the rise worldwide. In parts of the United States, home owners began spray painting messages to thwart potential robbers: "No copper, only TVC." Entrepreneurial types were melting down pennies minted before 1984. The value of the copper was far higher than the penny, leading the *Financial Times* to publish an article entitled "Melting Coins Could Start Making Cents." Engineers on Plan C were dismayed to discover thieves had broken into the Iquique storage facility and stripped the copper from the cables of their rig, removing the sophisticated circuitry. "It was a little frustrating to come back to Chile. . . . Some electrical cables were stolen," said Shaun Robstad, the lead engineer. "All the cords were gone, so my electrician got on the phone and started ordering cable. It was all put together in Houston. . . . A lot of people worked weekends and nights to get it done."

Day 38: Sunday, September 12

As dawn broke on Day 38, Golborne and the Codelco team began to consider the unthinkable: abandoning Plan B.

The original rescue plan had called for drilling three separate boreholes to the men:

one for the *palomas* with food and supplies, one for telecommunications, and a third for water and fresh air. Plan B had cannibalized one of the three original boreholes, forcing the rescue workers to combine the telecommunications with the water and enriched air. Now only two tubes were left. None of the engineers were willing to risk losing another borehole to accommodate the experimental plans of *los gringos* from Pennsylvania. If the tungsten drill could not be removed, a new shaft would have to be started from scratch. It would be a blind shot, a drill without a borehole to guide it down.

At the helm of Plan B, Jeff Hart was going stir-crazy. He had flown halfway around the world to help save the trapped miners, and now for the third consecutive day, the drilling was stalled. The fragments of metal drill bits at the bottom of the mine were wedged tight. Repeated attempts to haul, pry or drag them out had failed. Hart was frustrated. A clock ticked inside his head. Every day of delay meant another day of suffering for the thirty-three miners.

Meanwhile, André Sougarret worked to coordinate the installation of Plan C, the gigantic oil-drilling rig that was being assembled in record time. Instead of the normal eight weeks, Plan C was being installed in less than half that. Still, it felt torturously slow in the context of Operation San Lorenzo.

With time spiraling away from a solution, Igor cajoled his way into a brief afternoon audience with Golborne. The exhausted minister listened to the young engineer's description of the Spider and approved it immediately. The Spider was sent down; pressure from above forced the teeth in. Slowly the Spider was reeled back up. At the surface, a metalworker with a blowtorch cut into the Spider's cocoon, slicing away the teeth one by one. In a spray of sparks, he removed the final tooth and out rolled the Spider's catch: a tungsten hammerhead. The gathered engineers cheered. Plan B might still stand a chance of success. It wouldn't have to start over from scratch, without a guiding borehole, as had been feared. Still, Plan B had lost time — and the enemy wasn't just the mountain; every hour counted. Now the rescue workers were losing sleep and growing beards, too.

Deep below ground the miners sensed the chaos above. Every time a drill stopped, silence filled their world — a terrifying void that renewed doubts that they would ever be rescued.

NINE
TV REALITY

Day 41: Wednesday, September 15

With hot food, clean clothes, cots that raised them off the floor and a miniature projector that brought them TV and movies, the men migrated from the harsh edges of physical survival to a more nebulous state — the monotony of waiting with no clear end point. A pipe with fresh water delivered 106 quarts of fresh water a day. Four thousand cubic feet of cooled, fresh air were pumped down to the men each hour, but the temperature inside the mine did not budge, stuck at a sweltering 90 degrees Fahrenheit with 95 percent humidity.

Twenty days after the arrival of food, the men had a new problem: "We had no garbage before that — quite the contrary, we were looking for garbage," said Samuel Ávalos. Men filled barrels with rubbish, then used heavy machinery to dump the refuse at the lowest levels of the mine. The lack of proper bathroom facilities was also becoming more and more of a problem. A slight breeze from below began to fill their living quarters with the smell of stale urine. It became so unbearable that the men began to urinate in the empty plastic water bottles, screw the top back on, and then deposit the now refilled containers in the garbage barrel. The air improved dramatically.

While rescue engineers and psychologists worked overtime to keep the men occupied with tasks, the miners nonetheless began to slack off. Chores went unattended. Discipline was slipping away.

"What really screwed us up was the TV. When the TV arrived, it ruined the communications; it was a big problem," said Sepúlveda. "Some of the guys would just stare at it; they were hypnotized and watched it all day."

The miners began watching the nightly news and were starting to discover the global impact of their drama. Sepúlveda's wildly heralded narration of the first video had won him millions of fans above ground, but deep in the mine this adulation ignited a simmering brew of jealousy. In order to escape the pressure, Sepúlveda began to disappear from the main living quarters. For hours he would wander the tunnels, stopping to pray.

"When we lost control or humility, I would go alone into the dark," he said. "I would find my own spot. You have no idea what it is like to be alone in there. I felt at peace."

Fights and arguments erupted from the constant battle over which channel to watch. Urzúa called up and complained that the TV was "destroying the organization" and asked for the broadcasts to be limited to news, some soccer and the occasional movie.

Many of the rituals the men had developed during the grueling first seventeen days were now fraying. With food and comforts arriving daily from above ground, the bonds of solidarity that had kept the men alive in the direst moments began to break. "During the different shifts, the men would go around and check on those who were asleep. They would put their hand on the chest of every sleeping man to make sure he was breathing. Because of carbon monoxide in the mine, they wanted to make sure he was alive," said Pedro Gallo, the phone technician, who spoke with the men by telephone daily. "These were known as the 'Guardian Angels.' . . . They were vigilant in protecting the men who were asleep, but when the TV began, they stopped doing the rounds. . . .

They preferred to watch television."

Regular mail now reached the miners. Each man waited hopefully for a paloma delivery with his name on it and a letter inside. But it soon became clear that not all letters were being delivered in a timely manner. "There was no way to have a conversation; the answers were always four or five letters behind," said the miner Claudio Yañez.

Family members began to wonder about the fate of letters that rescue officials described as "lost." "Some of the letters were simply crumpled up and thrown out," said Dr. Romagnoli, who was clear that he did not approve of the measure. "That was done in the building where the psychologists were working."

The younger psychologists allegedly began to file letters with the Ministry of Health, in protest of what they saw as unethical censorship.

During phone calls with family members the miners accused the government of sabotaging family relationships. They began to fantasize about putting Iturra, the psychologist, in prison. "They asked me if there were any police up here at *la paloma* who could lock Iturra up. They said they would send the police rocks filled with gold, as a prize. I told them, 'Sure, consider it done,'" said Dr. Romagnoli as he described the miners' desperate attempt to eliminate Iturra from their lives. The miners believed their plan might work: "They sent up the rocks."

Frustration over the days of delayed and lost letters finally reached a head when Alex Vega, one of the quieter and more reserved miners, exploded as he spoke to Iturra about the censorship. In a fit of swearing and insults, Vega threatened Iturra, telling him that in order to communicate with his family, he would climb out of the mine. To his colleagues below, Vega described how he was going to attempt to scale a series of cracks and narrow chambers in the mountainside that the men were convinced weaved and rose to the surface. It was a mission they all recognized meant almost certain death. But, in the end, without professional climbing equipment, food and long-term lighting, not even Vega was willing to follow through on his threat.

Above ground, despite the battles, Iturra continued with his controversial system of rewards and punishments. "They should not have been given TV, that should have been traded for something," the bearded psychologist said with frustration in his voice. When the miners behaved well, they were given extra TV and mood music. Other treats — including live images of the topside world — were held in reserve. Should the miners deserve a reward or become unduly feisty, Iturra was ready with either a carrot or a stick. The miners began to rebel against what they saw as oppressive treatment. In a show of strength, they began to reject the daily psychological sessions.

When a set of personalized leather dice games was sent down, the men protested. Three of the games had typographical errors in their names. The men sent the dice and cups topside with an angry letter.

"The miners are like children," said Dr. Díaz, the lead physician who said that after satisfying their primal needs, including hunger, the miners were moving up the food chain of requirements. "Now that they have food, water and sustenance, they are asking for clothing, and we are seeing them rise to a third level: demanding that the food have an enjoyable taste. The other day they sent back dessert — the peaches — because one of them didn't like the taste."

In response, Iturra's team meted out more punishments. "When that happens, we have to say, okay, you don't want to speak with psychologists? Perfect. That day you get no

TV, there is no music — because we administer these things. And if they want magazines? Well, they have to speak to us. It is a daily arm wrestle," said Dr. Díaz. "NASA told us that we have to receive the arrows so that they don't start shooting the arrows at each other. So we are putting our chest forward; now they can target the doctors and psychologists."

Openly critical of what he saw as a provocative strategy, Dr. Figueroa, the psychiatrist hired to watch the operation, accused the mental health team of treating the miners like laboratory rats. First they try out unusual protocols, he said, then they study the results as if it were an experiment. "It's dangerous to implement psychological intervention without the consent of the miners," said Figueroa. "They are meddling in their lives. . . . This is an attack on the dignity of the miners. . . . The fact that they can hold out doesn't mean they are invincible or especially resistant. . . . They are very fragile."

Iturra was undeterred by the rising criticism. "We removed the first page of the newspaper and the miners went crazy and were screaming," he said, defending the censorship. The newspaper article described a mining accident in the Copiapó region where four mine workers had been blown to pieces by an accidental detonation of explosives. Iturra said, "One dead miner had the same last name as one of the miners down there; maybe he was a relative? I did not have time to check that out and we couldn't let them find out that way — by reading the paper."

"Disinformation and uncertainty are two of the worst psychological aggressions for humans," wrote Figueroa in a blistering critique of the San José psychological team. "Accurate, timely, honest and realistic information is essential. The benefit of restricting information delivery because of concerns about giving them bad news is not supported by empirical evidence and can compromise confidence in rescuers." But Figueroa also acknowledged that Iturra had a nearly impossible task. Miners were known to be among the groups least receptive to psychological counseling. They tended to hide their weaknesses, said Figueroa, who stressed the difficulty in bringing mental health to a group that was stubbornly opposed not just to Iturra but to everything he stood for.

At the miners' video conference with their families, the joy of face-to-face contact was now clouded by a bitter sensation that the psychologists were preventing any semblance of normal communications. While family members assured Victor Zamora that they had written him fifteen letters, he had received only one, and he began to think his family was hiding something. "Victor is very upset they are not delivering the letters," said Zamora's nephew. "He is about to explode. This is all so disgusting. None of the letters are arriving."

Inevitably the media began to question the censorship. During an interview with a Chilean broadcaster, Iturra defended the practice. "He said the opinion of the families did not matter, that the miners were 'his children,'" said Pedro Gallo, paraphrasing the conversation.

Later that same evening, a group of twelve miners gathered to watch the nightly news. As usual, Pedro Gallo, the telecommunications inventor, sent the video feed down to the trapped men. He was also sitting in front of a monitor, keeping tabs on the underground world. He was stunned by their reaction to Iturra's comments. "I saw their faces when the news came on and they heard Iturra. . . . Then the phone started ringing."

Irate, Sepúlveda called up and demanded to speak to Iturra, who had gone home for the day. Gallo knew that a battle royal was imminent. "Mario didn't say anything to me, but I could tell from his voice he was very upset."

Gallo explained that Iturra had broken a sacred miner code: he insulted the family.

The unity of the miners was now strained by different individual priorities, sleeping and group work shifts. The miners continued their daily get-togethers including prayer and the noontime group meeting, but fewer men participated. The necessities of survival were now being tempered by the relative comforts provided by the rescue team. In crucial moments, however, like the rejection of censorship, the miners still spoke with a single voice.

Day 42: Thursday, September 16

In the morning, Sepúlveda called and asked to speak to Iturra. It was an urgent request. Iturra came on the line. Gallo again was front row and knew fireworks were about to explode. Sepúlveda accused Iturra of abusing the rights of the miners. After a halfhearted defense, Iturra became quiet. Sepúlveda went on the attack once more. "If you keep shitting inside our home, we will remove you. This is your last chance," said Sepúlveda. He made it clear that he was going to report the incident to Mining Minister Golborne.

In a series of phone calls to political authorities throughout the rest of the day, the miners launched a counterattack. "He was treating us like little kids," said miner Alex Vega. "Of course we had to protest against the censorship."

Having just barely recovered their strength, the men now said they would not accept food or supplies. "We told them that if they didn't stop the censorship, we were not going to receive *la paloma* and would stop eating the food," said Barrios. "Everybody was against the psychologist; he did a terrible job. If he was not removed, we were not going to eat. We'd just leave the *palomas* filled with food," said Samuel Ávalos. "Like good miners, we pulled a strike."

After almost starving to death, the men were now threatening a hunger strike.

While the miners complained to the government, there was only so much the government could do. Iturra had been hired by the privately owned workers' health insurance company, ACHS. "We tried to fire Iturra, to dump him," said a high-ranking official in the Piñera government who asked not to be named. "But they threatened us, saying that if they did not control the psychological counseling, then they would not cover the medical parts of the rescue. We were trapped."

For his part Iturra called the battle cathartic. "I told them that I was going to be their father, if they want to get angry at me, get angry, but I am going to be their father and I will not abandon them, I am here and I am trustworthy."

Tensions with Iturra were becoming unmanageable. Dr. Díaz encouraged Iturra to take a break and suggested a week's sabbatical from the intense routines. After more than a month at the site, Iturra, straining under the pressure from the miners, lack of sleep, and the responsibility for maintaining the sanity of the thirty-three lives on his watch, agreed and went to his home in Caldera, a fishing port less than an hour's drive from the mine.

Claudio Ibañez, a psychologist from Santiago who had been assisting Iturra, took over the day-to-day counseling. With the miners rebellious and empowered, it was a tense time. With weeks — perhaps months — of captivity still ahead, it was crucial to keep the men healthy and calm. The rescue effort was dragging. The drills were advancing but hampered by technical difficulties.

Ibañez, an easygoing man with an extensive background in what he calls "positive psychology," upended the rules. There would be minimal censorship. *La paloma* would not be searched nor letters revised.

With the restrictions lifted, the families began to pack *la paloma* with secret gifts. For miner Samuel Ávalos, the change was a godsend. An avid reader, Ávalos was bored by the pamphlets from the Jehovah's Witnesses and the feel-good psychology texts sent by Iturra. Ávalos wanted drama, something so shocking it would transport him from the inside of the mine. "I read *El Tila,* a biography of a psychopath killer in La Dehesa [a wealthy neighborhood in Santiago]. It was fantastic. I read it three times," said Ávalos.

"I think this was a mistake. I was in favor of control," said Katty Valdivia, who is married to Mario Sepúlveda. "One woman snuck a letter to her secret lover inside the mine — one of the miners. She told him that she was pregnant," said Valdivia. "Then the wife found out and it was very tense for everybody. That kind of message should not have been sent down."

With the relaxation of the rules, more than just letters were making their way down to the miners. Valdivia described how the families "started inserting cigarettes, pills, even drugs into the paloma. It should not have been so free. Some miners became angry and bad feelings developed." Amphetamines were reportedly sent to the men. According to Valdivia, some authorities were aware of the contraband but turned a blind eye. Meanwhile, chaos ensued below. "The opening of the gates created conflict below in the mine, among the men," said Valdivia. "They went from strict controls to suddenly no controls."

"Before we noticed, the families managed to smuggle contraband down there," said Dr. Romagnoli. "The miners were not allowed to have candy because of all the dental problems, but still the families smuggled down chips, chocolate and candy."

Even a simple infection, like Zamora's inflamed tooth or a spike in insulin levels for Ojeda, the diabetic, could spiral quickly into a crisis. Doctors above ground were determined to avoid the most drastic scenario: having to guide Yonni Barrios in surgery. Yet the shadow of fear that Barrios might be forced to operate was always present. Delivery of unsanctioned food increased the odds.

Ávalos noticed that some of his colleagues were now acting suspicious. They were peeling away from the group in small cliques, wandering toward the bathroom — to smoke a joint, he suspected. "They never even offered me a toke," said Ávalos. "When you saw five of them headed up to the bathroom, you knew what they were doing." Ávalos was desperate for a quick hit, a high that would relieve the stress of nearly a month underground. "We went over to the area where the guys used bulldozers; we knew they smoked marijuana. They worked inside a plastic cab that protected them, and they could smoke a joint, then smoke a cigarette and no one would know. We looked everywhere for a *colilla* [stub of a marijuana cigarette]." They could not find one.

With group unity and long-term health key factors in the men's ability to survive, the temptations of short-term pleasures — alcohol, cocaine, marijuana — were in direct conflict with the needs of the group. Having small amounts of drugs circulating in the community created more tension than it relieved, instigated jealousies and threatened to alter basic tenets of the communal living conditions. Officials from the Chilean government became so concerned that they discussed putting a drug-sniffing dog at the paloma station. "We'll turn it into a border crossing," said one official, only half in jest.

But the men's greatest need would not fit down the tube: women. With physical health improving rapidly, sex became a topic of conversation both for the miners and for the rescue team. The men's sexual impulses were surging back, though they were still far from normal. "I am sure they put something in our food, something that kept us from thinking about sex," said Alex Vega. In fact, the medical team was working on another plan: how to appease the expected rise in sexual desires.

"There was a guy who offered inflatable dolls for the guys but he only had ten. I said thirty-three or none. Otherwise they would be fighting for inflatable dolls: Whose turn is it? Who was seen with whose fiancée? You are flirting with my inflatable doll," said Dr. Romagnoli. "It was supposed to be a relaxation tool. . . . The miners had a special place where sex with the doll could take place and they asked us to send four or five dolls and condoms. They could take turns. It was all planned. If we had thirty-three dolls there was no problem and each could do as he wanted with his doll . . . but I couldn't ask them to share."

The dolls were never sent; instead the men received pornography and pinup posters from *La Cuarta,* a Chilean tabloid famous for its girls known as *Bomba 4.* When the miners felt the need for privacy, they would block the government camera by sticking one of the pinups on the camera lens.

Day 44: Saturday, September 18

The Chilean Independence Day offered a welcome break from routines for rescuers, miners and families alike. This year September 18 was also the nation's much-anticipated bicentennial. Instead of battling over censored letters or plastic dolls, the miners agreed to a government-staged event: they would hold a holiday ceremony, eat special foods and sing the national anthem.

Chile's long-awaited bicentennial was overshadowed by Los 33. Above ground, Commander Navarro of the Chilean submarine fleet led a flag-raising ceremony on the flattened area used for the daily press conference in an effort to provide institutional symbolism to the historic date. Next to the flag, a banner with the men's faces had been strung up.

Two thousand three hundred feet below, a simple ceremony was under way. Omar Reygadas pulled a string that hoisted a small Chilean flag. Inside the tunnel, Reygadas raised the flag as high as possible, barely 3 feet above his head. Then Sepúlveda began his unique rendition of *la cueca*. With his miner helmet in one hand, a white towel in the other, Sepúlveda began to dance. He spun with the gusto and pizzazz of a *huaso,* a Chilean cowboy. A traditional *cueca* is a dance of courtship in which the macho man with a wide-brim hat slams elaborate silver spurs to the ground while the woman spins, sending her long hair and skirt into a flirtatious whirl. She skips aside, not to dodge the man's sexual offers but to encourage them. Sepúlveda's *cueca* was a solitary show.

The miners had built a small stage; a plastic orange tarp was hung on the wall and covered with a crudely written copy of their motto: *Estamos Bien En El Refugio los 33*. A Chilean flag was taped to the middle of the tarp and from the ceiling hung a series of patriotic buntings the colors of the Chilean flag: blue, white and red. On the edge of the tarp was written the names of the three groups: El Refugio, La Rampa, 105. The barren cave was now garish in the bright lights, like a crude theater set. In rubber-soled shoes, long white socks and hairy legs, Sepúlveda jitter stepped on the sharp rocks as his *compañeros* watched with obvious boredom. On the final stanza, he dropped to his knees, opened his arms like a devout pilgrim in joyous rapture and sent his energy skyward, through a half mile of solid rock. "Thank you all — our close colleagues who are out there working for us! We are grateful for what's been done and want to thank you." Sepúlveda's voice broke — a testament to the burdens of leadership. The camera panned the men — their faces impassive, showing few smiles and little interest. Instead of accident victims, the men were now beginning to feel like actors.

Day 46: Monday, September 20

In his daily press conference, Mining Minister Golborne was optimistic. "The three plans are advancing as expected," he said. Plan A — the first operation, which began on August 29 — had reached 1,066 feet, nearly halfway to the men. Both Plan B and Plan C, said Golborne, were now advancing at the rate of 3 feet an hour.

As the media circus at Camp Hope mutated into a full-fledged zoo, media expert Alejandro Pino was designing a strategy to help the miners cope with their newfound status as celebrities. Pino, a lanky sixty-seven-year-old with five decades' experience as a journalist and public speaker, conducted a five-hour class on media strategy for the men. The abbreviated course included interview techniques, marketing opportunities, how to handle tough questions and overall guidance on how to survive a pack of paparazzi.

As a longtime journalist and head of the regional division of the ACHS, Pino was neither paid nor obligated to provide media training but he felt a sense of responsibility for the men's welfare. He had a strong desire to help them with the oncoming onslaught of microphones and cameras.

Pino's early-afternoon classes were a welcome break from the technical talks about designing the rescue shaft or the much discredited psychological counseling. The miners gathered at their makeshift stage at the bottom of the mine. Pino, microphone in hand, worked out of a shipping container that had been outfitted with a white couch, some plants and a large TV screen on which he could see the miners down below.

Instead of warning the miners about the perils of media overexposure (as many people speculated), Pino went straight to the pocketbook. "If you do not look at the camera, if you give a boring interview, you will never be invited back for another interview," Pino told the men. "This is an opportunity, and you must learn to use body language, to be excited." An exuberant personality with a booming bass voice, Pino was practically evangelical in his effort to turn the shy and confused miners into media stars.

The miners gravitated to Pino's daily lectures. Though many remained invisible to the camera, they continually chimed in with questions, ideas and comments as their confidence and rapport with Pino grew. Given their animosity toward and distrust of Iturra, some of the miners refused to accept Pino, including Samuel Ávalos. "After the bullshit with the psychologist, we did not want to talk to those kinds of people," said Ávalos. "We did not like the idea."

Other miners were practically desperate to talk; they adopted Pino as their de facto psychologist. In the middle of a talk on media strategies, one of the older miners went off topic and began to confess to Pino: "If there is anything I have learned down here, it is that the last twenty years of my life have been a waste. . . . When I come up, I am getting divorced."

Divisions among the group also began to surge. Luis Urzúa was unhappy that some of the miners, including Victor Zamora, had acquired a video camera and were filming the others. And when a copy of *Ya* magazine arrived with an interview in which Sepúlveda bragged that he was "the leader" of the pack, further squabbles erupted.

"There were quarrels and little fights. They were getting very cocky and fighting verbally, but not fistfights. . . . No one lost their mind," said Dr. Romagnoli, who said *his* biggest brawls were in confrontations with the engineers topside. "I had problems with the guys on the surface. They did not understand the importance of the health care operation. They could have their Plan A, Plan B, Plan C, Plan whatever, but if the guys die? Then all those drilling plans are useless."

Day 47: Tuesday, September 21

For years Professor Nick Kanas had studied astronauts. He knows all too well the pattern of behavior when men are confined and stressed for long periods of time. In what he described as "third-quarter syndrome," confined men become increasingly anxious and irritable toward the end of a mission — in this case the rescue day. "After six weeks, people tend to get territorial. There is often not a lot of joking and banter, although they try. They will start to form subgroups," said Kanas, who works at the University of California, San Francisco, and has been a longtime consultant to NASA. “After six weeks, the situation turns sterile and confining. What was once quirky and fun — like the jokes of a colleague — becomes irritable and tiresome.”

The men had now been underground for forty-seven days. Whether it was footage of a miner unscrewing the *paloma* and pulling out food or footage taken as they raised the flag and sang the national anthem, their entrapment was being captured on video. TV reporters repeatedly attempted to smuggle cameras down the *paloma* so the miners could start producing an underground documentary.

When Chilean detectives working for the Polícia de Investigaciones de Chile (PDI) needed evidence to document details in the criminal case against the mine owners, they taught the miners the basics of crime scene photography. For a week the miners were like stars in *CSI* as they documented and filmed faulty security inside the mine. Florencio Ávalos went to the far corners of the mine to film cracked walls, rusty piping and the huge boulders strewn across what was once the main road inside the mine. An estimated forty hours of criminal evidence was filmed by the miners and then shipped above ground.

Day 48: Wednesday, September 22 — The Drill Falls Down

On September 22, the miners received a surprise delivery from above. When Plan B reached 280 feet, the bit snapped and one of the four drill heads went into free fall the length of the shaft and fell into the floor of the mine. No one was injured as the metal hammer plowed into the mud, but Plan B was halted.

Mario Sepúlveda called the rescuers. "Ah, I think we have something of yours down here," he said in jest. "I believe that it is called a bit, a drill bit. But what is it doing here?"

Juan Illanes, one of the miners, hauled the bit out of the mud. The miners had vast experience with heavy machinery; the reality they lived every day was constantly fraught with broken parts, last-minute improvisations and setbacks. But this was unbearable. The frustrations began to boil up. "They are working to rescue you and you have this kind of failure? It was depressing," said Samuel Ávalos. "It means two more days — five more days. We were receiving food from *la paloma,* but we were confined. Trapped! That was killing us."

Day 49: Thursday, September 23

The tensions below continued to rise.

Ibañez, the psychologist, had a cordial relationship with the miners, and many of the men liked his relaxed and positive attitude. However, Ibañez was unable to keep control. The breakdown of authority, the constant TV, the drugs being smuggled in — all of this was seen by many miners as a grave error. The men stopped listening to advice from above ground and began inventing their own activities.

Edison Peña began to explore the tunnels. Before the accident, he had been an exercise enthusiast, riding his bike for an hour every day, followed by a long run. Now he began to jog a 3-mile circuit inside the tunnels. His stiff, above-the-ankle mining boots chafed his legs so he took a pair of wire cutters and chopped the boots down. The sharp rocks and uneven terrain aggravated his already damaged knee, but Peña kept running, as if he might escape either the terror of the tunnels or the nightmares in his head. Pablo Rojas, Mario Sepúlveda, Franklin Lobos and Carlos Mamani joined Peña. The men huffed and sweated then rested in the deepest part of the mine. With their white plastic protective suits, hoods and goggles, they looked like astronauts.

Games and books began to arrive via *la paloma.* Inside the refuge, the men organized marathon sessions of dominoes and card games. In many ways the cards were just an excuse for the ongoing banter and constant joking, monologues and double-entendre word games. Among the miners, the ability to talk down an opponent — at the card table or in the daily meeting — was fundamental in establishing respect. While arguments raged, fistfights were extremely rare — some miners say nonexistent.

"I had to club Ariel [Ticona] in the head with my head lamp," admitted Sepúlveda, who stressed that the incident was an isolated case of physical violence among the men. Ticona had allegedly insulted Sepúlveda's mother, explained one miner who witnessed the fight. "If we had let the violence disintegrate to that level, we would have ended up with a few guys dead."

When Ibañez attempted to bring a reporter from a Chilean TV station into the video conference room to interview the miners, another firestorm erupted. The reporter was escorted into the small container high on the hill, an area off limits to anyone without a Rescue Team pass. Ibañez asked the miners if the journalist could conduct a few interviews with the trapped men. Urzúa, Sepúlveda and the other men went ballistic. Immediately, the phone above ground started ringing over and over. The miners were indignant. Gallo listened to the ensuing conversation.

Urzúa berated Ibañez, saying, "Hey, stop screwing around with us, okay? Why are you sticking a journalist in here? We don't want any journalists here! We don't want to be seen. We are suffering down here, so no interviews, and we are going to report this. We will report this."

The miners called Rene Aguilar, the number-two man on the rescue team and Sougarret's right-hand aide. A psychologist by profession and a ranking executive with Codelco, Aguilar had been in the front row for weeks. Now he was furious. "He came over to speak with Ibañez and his face was bright red. He was angry," said Gallo, who witnessed the scene. "And that was the last time I saw Ibañez. He went to Santiago. We never saw him again."

With Ibañez gone, the miners were quickly reunited with their former psychologist, Iturra. "This was Iturra version two-point-oh, the diet version," said Gallo, whose job included monitoring the live video feed of the miners below. "He came in with a totally different attitude and offered his two-hour-a-day slot for psychological counseling to the men. He said they could use that time to speak with their families."

With the miners now holding the upper hand on censorship, the missing censored letters suddenly appeared. "It was like a rain of letters. They sent down all of them — I would say three hundred letters, all at once," said Gallo.

TEN
FINISH LINE IN SIGHT

Day 50: Friday, September 24

On September 24, the miners spent their fiftieth day trapped underground. In the centuries-long annals of mining, no miners had ever been trapped underground for that long and survived. No one at Camp Hope celebrated the grim record. But instead of despair, there was now a glimmer of light. The families could imagine the rescue. Three separate multimillion-dollar drilling operations were pounding toward their loved ones. Food was being delivered every few hours. The mail service was still suffering from delays and occasional censorship but at least it was operational, and the laundry service impeccable, though everything still had to be rolled and squished to fit inside the narrow *paloma*. Many of the wives personally washed and ironed their husbands' dirty clothes. Sometimes they doused the shirts with their favorite perfume, in anticipation of a forthcoming sexual reunion.

Dreams of being united were further stoked on the afternoon of September 25 with the arrival of the Phoenix. The missile-shaped rescue capsule — custom-built for the San José rescue operation by the Chilean Navy using specifications from NASA and the successful Pennsylvania rescue operation at Quecreek — was painted the colors of the Chilean flag. Weighing in at 924 pounds, with an interior chamber 6.5 feet high, the rescue capsule became a character unto itself. Ministers and rescue workers posed inside the circular tube. Family members approached the contraption and touched the capsule lightly, as if it were a sacred totem.

While the rescue plans advanced without a hitch, Dr. Romagnoli, who had gained the miners' confidence through his handling of the Iturra and Ibañez situations, now began preparing the men for the escape mission. Romagnoli knew that if the capsule failed or became stuck, the rescuers might be forced to haul the men out by a far simpler and riskier method: strapped and tied to the end of a long cable. But no matter what, the men needed to be in the best shape possible — they might be forced to climb ladders, descend ropes or, if the capsule jammed, simply stand for an hour in the tightly confined space.

Romagnoli, an adviser to both the Chilean armed forces and professional athletes, began teaching the men light exercises in preparation for more strenuous physical education classes. He recommended that the men jog as a group in a one-and-a-quarter-mile stretch of tunnel. Using the U.S. Army fitness training as a model, the men sang while they jogged. Romagnoli explained that singing was a safety precaution to keep their heart rates in check: "If their heart rate goes above one-forty, they can't sing and jog at the same time."

Romagnoli said the men were enthusiastic about the new routines. "One of the advantages we have is that these guys are strong; they are accustomed to working their arms and upper body. This is not a sedentary population we are dealing with. They will respond quickly."

Using a sophisticated chest-mounted device known as the BioHarness, Romagnoli harvested a wealth of data on the trapped men. Now the miners provided real-life data on extreme situations to the experts at NASA. "The Chileans are basically writing the book on how to rescue this many people, this deep, after this long underground," said

Michael Duncan, one of the NASA experts who visited Chile.

In addition to his deft handling of the psychologists, Romagnoli won over the miners early on when he supported their pleas for cigarettes. He was a smoker and had openly questioned whether the most stressful moment in the miners' lives was an appropriate time to ask them to kick the nicotine habit. Romagnoli was unorthodox: he believed in commonsense solutions, even at the expense of textbook wisdom.

Sitting behind his desk, high on the mountain at the *paloma* station, Romagnoli's daily chores of shipping medicine, logging vital signs and chatting with the miners consumed just a fraction of his twelve-hour shift. Now that the men were relatively comfortable, instead of hearing about urgent health problems, Romagnoli was pestered about the most minimal life improvements. One letter from below was a complaint that the miners had run out of artificial sweetener. Another miner sent Romagnoli his MP3 player and complained that he had too much reggaeton and not enough cumbia music. Romagnoli began downloading music, erasing and then reformatting the MP3 player with a customized song list. "These guys aren't sick anymore," he said with a laugh. "Now they think this is room service and I am their fucking DJ."

Day 52: Sunday, September 26

As the miners continued to send up videos from below, the front-row characters, including Sepúlveda and Urzúa, were becoming known worldwide as the charismatic cheerleader and the rock-solid shift foreman, respectively. Many of the other miners remained anonymous. Not only were they not visible in the videos, but they were invisible when it came time to work. A division had begun to form between those who were willing to volunteer and advance the rescue effort and those who lazed around waiting to be saved. Despite massive efforts to keep the men busy, their life was now reduced to killing time. It was exactly the situation that NASA had warned against — free time in a stressful and barely habitable environment was an incubator for problems.

Arguments broke out between the men who worked and performed duties and those who did not. A half dozen men lay in bed, staring at the rock ceiling, listening to music on personal stereos, or lounging uncomfortably in the TV room. "They were lazy, they did nothing," said Franklin Lobos, describing the attitude of several of the miners.

First it was the arrival of television that had been a distraction for the men; now boredom and a relative sense of security were threatening the harmony of the group. Samuel Ávalos — whose official duty was to measure the daily temperature, humidity and levels of potential fatal gases inside the mine — said his job was a daily exercise in monotony. "The temperature never changed; it was always around thirty-two degrees [90 degrees Fahrenheit] and the humidity at ninety-five percent," he said while recounting how the heat was driving the men mad. Victor Segovia, the tireless scribe, began to have nightmares that he was trapped in an oven.

Day 54: Tuesday, September 28

After weeks of technical setbacks, all three drilling operations were now grinding slowly toward the men. Plan C was finally up and running. One hundred sixty-four feet high, the massive platform towered above its "competition." Using huge drilling bits that looked like dinosaur claws with gear, Plan C was the talk of the engineers. They estimated that within twenty days, the petroleum rig would reach the men. Bets were placed on what day the rig would break through to the miners down below. It would be another week before they all realized that the rock at the San José mine was so dense — more than twice as hard as granite — that the rugged petroleum rig would advance far slower than the engineers had hoped and the miners imagined.

Sougarret was now facing a delicate decision. Would each tunnel need to be reinforced with steel casing? The advantage of the casing was the ability to provide a uniform surface for the retractable, adjustable wheels of the Phoenix capsule. No one dared imagine the logistics of the Phoenix getting jammed in the tube and having to organize a rescue of the rescue workers or, worse, of a nearly freed miner. All efforts to make the last ride up uneventful were being thoroughly explored. Sougarret, though, knew that installing the tubing would add another three to seven days to the entire rescue operation. The weight of the tubes was estimated at 400 tons, meaning a special crane would have to be driven in from Santiago to install them. Repeated inspections of the carved tunnels showed a near-perfect surface — glassy and like marble in many sections. However, the first 328 feet were far less uniform and prone to crumbling or disintegration. Sougarret refused to make the final evaluation; for now his eyes were focused on reaching the men. The pressure of maintaining the men's physical and mental health remained.

Day 55: Wednesday, September 29

At Camp Hope, the level of action was also agonizingly slow; thousands of journalists were going stir-crazy. Access to the actual rescue operation was limited to government cameras and a few fortunate journalists who had been given insider access — including those from the Discovery Channel, a Chilean documentary crew, and the author of this book.

In response to the media's hunger for images and footage, the Piñera press team set up a command post at the mouth of the San José mine. Piñera's aides working in a unit known as the "Secretariat of Communications" reviewed tapes to determine if they were suitable for public release. Brief snippets were released to the press, but hundreds of hours were never shown, as government lawyers began to debate the legality of airing more footage. If the mine was effectively their home, what were the rights of the miners with regard to the videos being filmed by government cameras? Was this rescue public or was life in the tunnels a private scene? Could broadcasting the videos lead to lawsuits against the government for invasion of privacy?

As the Piñera presidency battled with privacy rights, the world's media continued to swarm into Camp Hope on a scale never before seen in Chile and rarely seen anywhere in the world. The number of registered journalists surpassed two thousand. For several acres, the rocky hillside near the mine was blanketed in motor homes, tents, satellite dish rigs, temporary plywood broadcast platforms and, increasingly, the cream of the world press. Photographers began to chain their tripods to key spots in an effort to guarantee a clean shot of the drilling rigs. TV crews argued over who had first claimed a towering boulder that served as a base for satellite transmissions. Every day a parade of new faces arrived at San José, lugging tripods, struggling with new phone codes, and gawking at the surreal scene.

Just beyond the crowded site near the mine, the desert was empty in every direction. Not a tree on the horizon, just sweeping golden dunes interrupted by the occasional tracks left by motocross racers from the legendary Dakar to Paris marathon. After being chased off the African continent by a combination of political instability, security fears and the tragic death of a pedestrian squashed by the careening crowd of foreign motorists and cyclists, in 2009 the race was transferred to this remote corner of Chile. Hundreds of journalists had covered the race and camped in the nearby hills. Now they were back to cover another competition: the race against time.

Photographers began to brawl and shove. With so many cameras and microphones, obtaining a clean shot was nearly impossible. The dust wreaked havoc on expensive lenses. And worse, the best shots had all been taken a thousand times over. One local newspaper dubbed the entire scene "the Woodstock of Media." A building war erupted when TVN, the Chilean national broadcaster, built a broadcast platform directly in front of the one used by CNN Chile, forcing the latter to add a story to its own construction. Ramon Vergara, a local carpenter who made a business off the broadcasters' competition, was cashing in. Vergara built three platforms in as many days. "I charge one hundred twenty thousand pesos per platform [$250]," Vergara told *The Clinic*. "I try to do one a day."

While the miners were reported to be in good health, the ACHS ambulance stationed high on the hill as part of the miner rescue was now regularly seen wailing down the hill, on a mission to rescue injured journalists. Ten separate accidents involving journalists and cars had been reported.

With a troupe of clowns and roaming Franciscan monks in robes, the scene began to feel like a circus. "It lacks only the lions," said Vinka Ticona, a relative of the trapped miner Ariel Ticona. Children playing in superhero costumes were so common that it was no longer strange to see a flock of little kids dressed as Spider-Man climbing the rocks like monkeys.

Nightly bonfires became the hub of the now eclectic friendships between journalists, policemen, politicians and family members. Isabel Allende, daughter of Chile's late leader Salvador Allende, could be seen giving an interview to CNN one moment, then sharing a fish sandwich and chatting with Isabel Allende, the Chilean novelist and a distant cousin of the former president. Long lines snaked away from a stand that delivered fish tacos. Free grilled seafood, homemade soups and a truckload of cookies kept everyone fed. Officially, the mine area was an alcohol-free zone. But the early-morning piles of empty beer, wine and pisco bottles were abundant evidence that the area was only alcohol free because it had all been consumed.

■ ■ ■ ■

Worldwide, millions of viewers were obsessed with the story line: Would the men make it out? Who would be first? The story was now a combination of Reality TV and live disaster, edited and fed to the media with the invisible but slick editing of Piñera's communications team. Instead of fulfilling expectations of chaos, violence or a *Lord of the Flies*–style disintegration, the miners were offering a rare moment of global unity focused on joy, hope and solidarity. The traditional TV slogan "If it bleeds, it leads" was temporarily upended by a nonviolent drama featuring a cast of underdogs.

At Camp Hope, an early wave of talent scouts and TV producers began to battle for the life rights to the miners' story, in particular to a 150-page diary kept by the miner Victor Segovia, who had been chronicling the daily activities, including the darkest moments of the seventeen days without food. Segovia's family began negotiating with publishing houses. The starting price for the unique memoir was $25,000. Tabloid reporters began sleuthing for the first exclusive interview with a miner, as they began signing up families and offering promises to whisk the men away on junkets to Los Angeles and Madrid.

Though the men were still trapped, a movie about their experience was already in production. In a nearby abandoned mine, Chilean and Mexican actors were reenacting the drama, adding more than a few touches of literary license, as many details of the men's day-to-day routines remained a mystery. Chilean director Leonardo Barrera also announced plans to film a porn movie based on the miners trapped underground. Barrera claimed his film would not be "a massive orgy" but a sympathetic and fictionalized account of miners having sex with *minas,* the Chilean slang for a sexy woman. The miners were about to be hauled from the dark, damp world of mining and tossed into the klieg lights of Hollywood. There would be virtually no time for a transition.

Day 57: Friday, October 1

On September 30, Edgardo Reinoso, the miners' lawyer, filed a suit against the government, seeking $27 million in damages and alleging government negligence in the reopening and continued operation of the San José mine. Reinoso was now representing all but three of the families. A month earlier he had successfully embargoed a quarter-million-dollar payment due to the

San Esteban mining company. Originally recruited by the mayor of Caldera, a coastal city near the mine, Reinoso hoped to have that money turned over to the miners, a part of what he saw as a multimillion-dollar settlement due the men.

The rotund showboat attorney was famous for winning a 2007 lawsuit against a local municipality when a pedestrian overpass collapsed, killing two spectators at a New Year's Eve celebration in the coastal city of Valparaíso. An avowed opponent of Piñera and the Chilean right, he was now determined to get money from the government. "We, the families, want them to pay for all the damages, and we want justice," said Katty Valdivia, Mario Sepúlveda's wife, who supported the lawsuit.

Reinoso's attack on the government was seen as a cheap shot by Piñera's aides, who never failed to mention that the dangers of the San José mine had been known for a decade and that El Concertación, the progressive center-left coalition that ruled Chile from 1990 to 2010, had done little to protect the workers and, in fact, repeatedly allowed the dangerous mine to avoid a permanent closure.

Opinion polls showed the new president's approval rating — the rather cheap and unreliable barometer of success in the contemporary political arena — had jumped from 46 percent pre-disaster to 56 percent as the rescue progressed. In August, President Piñera had placed his credibility on the front lines of the miner rescue. Now, with Reinoso's lawsuit, he risked losing that huge capital gain.

President Piñera was also coming under criticism at the mine, where some rescue workers were appalled by the president's actions. They accused him of using the rescue for political advantage. Dr. Díaz, the lead doctor for ACHS, criticized Piñera and Golborne for altering medical and technical protocol to steal the spotlight. "These guys want to be in front of the cameras as the great saviors," he said, frustrated that the rescue operation was being compromised by staged PR moments that, as he saw it, were designed to benefit the president. "At some point it is going to be pretty difficult to bite my tongue."

In an article on CNN's Latin American page titled "Family Members Accuse Chilean President of Using Them," Nelly Bugueño, mother of the trapped miner Victor Zamora, denounced Piñera. "This is all politics. It is dirty. It is a fraud and propaganda," she said. "They are playing with the sentiments of our dear families."

Other family members acknowledged that they had no love for Piñera or his politics but that his government had done everything possible to save the miners. "Personally, I can't stand the guy and have vast differences of opinion. But he's made great decisions," said Cristian Herrera, nephew of the trapped miner Daniel Herrera. "If you ask me do I have to thank him? Yes. If the previous government had been in charge, the miners would have died."

On October 1, Minister Golborne ended a month of rumors and speculation by confirming what was by then an open secret: the rescue effort was advancing far faster than publicly acknowledged. "The good news is that thanks to an analysis that we have done together with the technical team, we can estimate that the rescue of our miners will happen in the second half of October." Golborne noted that the drills had passed through the loose top layer of rock and were now in a geologically more solid section of the mountain. "This allows us to be slightly more optimistic," said Golborne, who also announced that he had already informed the miners of the same good news.

With the rescue of the miners now speeding ahead, deeper questions were beginning to be asked and a broader debate erupted in Chile. Why were the miners trapped in the first place? Why was such a notoriously dangerous mine still operating? A congressional investigation in Chile begun in late August had unearthed a damning history of fatal accidents at mines owned by San Esteban, the holding company that owns the San José mine as well as San Antonio, an adjacent mine.

Figures provided by ACHS to the Chilean congress showed that the rate of accidents at the San José mine was 307 percent higher than the industry average. "The average company pays 1.65 percent of a worker's salary for insurance; they paid 5.37 percent," testified Martín Fruns of ACHS, who noted that the San José mine owners had not paid the workers' insurance fund for five months.

In testimony before the committee, María Ester Feres, a former head of the government's labor ministry, said she had tried to shut down the San José mine nearly a decade earlier, in 2001, but was rebuffed by what she described as "pressures from the mining sector" and concerns that those jobs would be lost. "There was some minor work done in the mine but the perception at the labor ministry was that this mine was a bomb . . . and it had no escape exits."

The congressional investigation also exposed the fact that inside the San José mine a steady rain of rocks was constantly crushing workers. Some were minor accidents that required no hospitalizations. Others ended in funerals.

Mine owner Alejandro Bohn testified that safety improvements were a "sacred principle of our company." Asked about the accident that sliced off the leg of worker Gino Cortés, Bohn blamed workers for not replacing a safety net designed to catch falling rocks from the ceiling. He then went on to say, "Unfortunately, it is the same shift that is now trapped inside the mine."

Many observers were shocked by Bohn's callous remarks. It sounded like blaming a lightning strike victim for not walking around in rubber-soled shoes. "The whole roof was without mesh — maybe twenty percent had mesh," said Samuel Ávalos, who was indignant when he heard Bohn's analysis. "Where were we supposed to walk? Where?"

Piñera's newly minted secretary of labor, Camila Merino, admitted that the Piñera government had been aware of the dangerous work environment. "We had indications of safety problems and we should have acted with anticipation. That is why it is important that all the safety measures that we are now proposing are taken notice of, so that we don't have more accidents in the future," said Merino.

Her comments caused a furor in Chile. Opposition lawmakers demanded more details. Had the government covered up gross safety problems at San José? If so, could they keep the lid on the scandal? Merino backtracked, insisting she had no solid information.

Javier Castillo, a union leader in Copiapó who had battled with the mine owners and with the national mine inspection service known as Sernageomin for more than a decade, was ecstatic about the newfound interest in worker safety. In hundreds of documents filed with the courts, local politicians and mine owners, Castillo had warned that both San José and San Antonio were frighteningly dangerous and on the verge of collapse.

While the world wondered why the San José mine had collapsed, Castillo was determined to show that government oversight had also collapsed. A video made in 2002 by the miners' union highlighted unsafe mine practices and the probability of a cave-in at both mines. In documents that he provided to the congressional investigators, Castillo showed that the owners of the San José mine had been warned that the mine was dangerously fragile. In 2003 the San Antonio mine — located on the same mountain as San José — experienced a massive collapse. Then in 2007, the San Antonio caved in again and was shuttered. There were no casualties only because the collapse happened at 1 AM when there were no workers in the mine.

The abundant details about a series of fatal accidents that Castillo provided led government safety officials to shut down the San José mine for all of 2007 and part of 2008. Now the congressional investigation was focused on a central issue: Should the mine ever have been reopened?

Under Chilean law, the San José mine was required to have two separate exits: the regular daily route and a backup for emergencies. After investigating, the Chilean congress concluded that at the San José mine, there was never a backup exit, and even provisional requirements like stairs inside the ventilation shafts were never implemented.

In the weeks and months before the final collapse, the San José mine had given signals of instability. In June 2010, a block of rock fell, smashing Jorge Galleguillos in the back. An investigation by the ACHS, the workplace insurance company, warned of the risk of further collapse. Alejandro Pino of the ACHS said the mine owners had been advised about the imminent dangers. He said, "We asked the company to firm up the mine."

Day 59: Sunday, October 3

As the congressional investigation continued, the rescue drew ever closer. The hillside around the San José mine was crawling with construction crews building a helipad, a temporary hospital and bleachers for the journalists. The government was even designing lounges where family members could sit on smartly designed couches with flowers, neon blue lighting and stylish hallways. All for their first, brief visit with the rescued miners.

With the miners now media-trained and gleaming with the imprimatur of freshly minted celebrities, long-lost relatives began to arrive at Camp Hope. So many unknown "family members" arrived that when the Chilean newspaper *The Clinic* published a map of Camp Hope, it included an arrow pointing to a section entitled "Family Members" and a second arrow indicating the campground for "Supposed Family Members."

Psychologists at Camp Hope scrambled to

prepare the families for the unknown consequences of the trauma. Would the men be joyous or subdued? Would they proclaim everlasting love to their wives or seek imminent divorces? And in the case of Yonni Barrios, the de facto doctor below, would he stay with his longtime lover or his wife? Many of the miners were thought to be suffering from depression. What would the long-term effects of the unique trauma be?

After his month-long battle with the men, Iturra was now more of a concierge and cheerleader. He avoided fights with the men and instead became a diplomat, soothing family problems, delivering messages and repeating his mantra that they were "one day closer to rescue" as he sought to hold the miners together long enough to get them out.

Even as Iturra worried over the men's fragile states of mind, Sougarret and his team were facing major obstacles.

A new round of technical setbacks stalled Plan A. As technicians rushed to change the hammers and head of the drill another three days were lost. Though Plan A was now less than 328 feet from the men, few engineers bet that the much-hyped original rescue plan would win the race. The slow but steady original drilling operation was now headed for third place. Both the other drills had proven faster in the unique conditions of the San José mine.

Plan C also faced a setback when an errant drill led the shaft far off track. Using a smaller drill bit, engineers devised a plan to curve the tunnel back on course, then return to drilling with the full-sized bit, which would bore a tunnel wide enough for the Phoenix. In total, nearly a week would be lost. But the speed of Plan C was now compromised by the inability to keep the massive rig on course.

All bets were now on Plan B, which by day 59 had reached 1,400 feet and appeared to be the most reliable technology in the epic operation. With Plan A and Plan C facing major challenges, President Piñera's decision to operate three separate technologies now seemed utterly prescient.

ELEVEN
THE FINAL DAYS

Day 62: Wednesday, October 6

With the Plan B drill less than 160 feet away, the men could hear the grinding and pounding so close by it seemed that at any moment the drill would break through the roof of the workshop. Or was this another deception? Rumors flooded the tunnel: the drill would arrive in one day. Or eight.

For the miners, food suddenly became less important than their daily diet of information. "At a quarter to nine you would hear someone banging on a can. It sounded like a bell and they would be yelling, 'News in ten minutes. News in ten minutes!' We all came to watch," said Samuel Ávalos.

The nightly 9 PM newscast was now the focal point of the day. In their underground studio, sweating profusely, stripped to white shorts and rubber-soled shoes, the men gathered to watch and listen to the latest developments. The mine rescue blanketed the coverage; the first twenty minutes would be dedicated to Operation San Lorenzo, the mine rescue mission.

"Minute by minute we knew what was going on," explained miner Samuel Ávalos. "We followed the rescue and started to calculate when the rescue would be over. We were too informed; that caused in each of us an anxiety of 'Let's get this over with. Get me out of here.' . . . Without so much information we would not have known when we were supposed to be out."

Rescue officials, including Sougarret, refused to provide the miners with a specific date. Golborne had ordered caution and Sougarret agreed. Many things could still go drastically wrong. In drilling the tunnel, the machines had occasionally drifted briefly off course and then been corrected and as a result the tunnel had slight curves and dips; would the capsule get stuck? One curve in particular near the bottom of the shaft worried the engineers. Rumors abounded at the command center that the Phoenix just barely fit along the curved section. Also, a small dynamite charge would have to be carefully calibrated to expand the spot where the Phoenix would enter the roof of the workshop. This, too, caused sleepless nights among the rescuers. If too much explosive was used, they risked collapsing the shaft in on itself. No one could predict how the walls of the tunnel would hold up to multiple scraping, sliding journeys of the Phoenix. On camera, the shaft appeared to be as solid as marble, but there was no way to know until the capsule was in operation. An earthquake was another terrifying possibility. Chile had been ground zero for two of the world's five biggest earthquakes, including the February 2010 earthquake that was still a fresh memory.

And as the whole world now well knew, due to the chaotic removal of gold and copper, the entire hillside surrounding the San José mine was fragile, a hollowed-out skeleton, propped up by the remnants of the mountain. Geologists had been clear in their diagnosis: the mountain was fragile.

As part of the rescue, Codelco had studded the hillside with sensors capable of measuring the slightest geologic movements. If another collapse was coming, the engineers hoped to have at least a brief forewarning.

For the trapped men, the message was contradictory. The rescue was imminent but indefinite. "Nobody could sleep. We were all so nervous," said Ávalos, describing the rising tension. "There was so much noise, the machinery going this way and that. Everyone was uneasy. Physically we were desperate. It was worse than the first days."

One measure of the rising tension was the number of cigarettes being requested from below. Instead of nine smokers there were now eighteen. Instead of two to four cigarettes a day, the men were given practically unlimited access to tobacco. When tobacco became short and the men got to the bottom of their allotted stash, tensions shot up. Fights nearly broke out.

Prescription sedatives also were being sent below. For some men, the drugs were an aid to help them sleep. For others they were used to slow down the excess adrenaline rush. In a select few cases, the drugs were a drastic move to stop what was looking like the onset of mild psychosis. Though it was never publicly acknowledged, in private health meetings among doctors and paramedics, the mental health of the individual miners was described in terms that included bipolar, manic depressive and suicidal.

To stave off the rising anxiety, the men developed an ingenious form of recreation. The drilling operations were lubricated with a steady stream of water. Using a crude system of canals, the men engineered a path that channeled the water away from their living quarters and down to the lowest levels of the mine. At first, the collected water was a muddy slop that was practically useless as a bathing spot. More mud was smeared on the body than removed. However, as the rescue effort advanced and the canals improved, the bottom of the mine began to fill with water. Eventually the pool measured 23 feet by 10 feet and was over 3 feet deep. By early October, the men had nicknamed the growing pool *La Playa* ("The Beach") and had enough water to swim and frolic. "I would swim laps," said Mario Sepúlveda. "We had a great time."

For hours the miners began to lounge in the pool, floating and laughing.

Pedro Cortés, an expert at driving the Manitou, a mining truck with an adjustable hydraulic platform, would drive the vehicle down to the pool, flip on the headlights and light up what became a surreal scene: a half dozen naked men, 2,300 feet below ground, carousing in a pool.

For brief moments the men could forget their tragedy. They told jokes, imagined a life of freedom above ground, and promised one another never to abandon their unique brotherhood. None of the men doubted they would sacrifice their life for another. Even the most strained relations had a core fraternal bond. "I could look into his eyes and know exactly what he was thinking; sometimes it was not even necessary to speak," said Samuel Ávalos, describing his connection to Mario Sepúlveda.

The men had a loyalty forged on the sharp edge of starvation and death. They had been condemned to die together, not in a sharp instant but over the agonizing course of days. "At the time we did not talk about cannibalism as a group," admitted Richard Villarroel. "Afterwards it was mentioned a lot in jokes."

The jokes about eating one another were a barely disguised admission of how close they had come to a savage and barbaric end. The emergence of poets, promises and compulsive joggers can be seen as ferocious attempts by the miners to regain their humanity, to shove the somber shadow of barbarism and death to an ever greater distance.

Day 63: Thursday, October 7

During their two-month captivity, the miners accumulated a massive quantity of gifts, including photos of naked women, miniature Bibles, hundreds of letters, fresh clothing and the occasional smuggled chocolate.

Each man designed a crudely decorated sleeping area. The men attached mesh netting to the rocky wall and hung the Chilean flag, family photos, letters and drawings. "I had a special area with both the God and the Devil," said Ávalos. "I had Luli [a busty blond Chilean pinup girl] with that great ass of hers and I had Mother Teresa of Calcutta. Those are my idols. Those were my inspirations." As he looked around his makeshift "room," Samuel Ávalos realized he was living like a pack rat. With no furniture or shelving, he hoarded his belongings on the rocky, damp floor of the tunnel.

Even at the depths of the mine, the trappings of celebrity were now apparent. Every other paloma now included flags and requests for all thirty-three men to sign and return them as soon as possible — flags from the Universidad Católica soccer club, the Cobresal soccer club, from Geotec (the drilling company) and, most of all, dozens and dozens of Chilean flags. The men obliged. Signing autographs and getting writer's cramp at 2,300 feet was a brief foreshadowing of the media swarm and fame that awaited above. But in their innocence, many of the men were unable to understand the dimensions of the world's fascination with their underground world.

With the drilling operation now on track, the men began to think about their move to the surface. What would they bring with them? What would they leave behind? Time — once their eternal enemy — was running out. It was now time to pack up.

As the days passed, the miners began to reverse the delivery system. Now it was a flood of items shooting up the *paloma,* a seemingly endless stream of paraphernalia including rock collections, diaries, flags and soccer jerseys signed by European stars, including the Spanish World Cup hero, striker David Villa, whose father and grandfather were miners.

At least twice a day, sometimes more frequently, Luis Urzúa was informed of the advances, setbacks and protocol for the rescue. The entire operation required massive and continual feedback from below. Sometimes the request was as simple as moving a camera to feed a live shot to rescue planners above. Other circumstances demanded that the men move heavy equipment into place to reinforce a weak roof, cut away loose rock or repair damaged communications equipment.

The miners began moving hundreds of empty water bottles, plastic wrap from food and wrecked equipment to a dump they created at the bottom of the mine. Muddy floors and swirling dust and dirt from the machines made cleanliness impossible; still the men attempted to leave their living areas organized. "It is like going on a trip," Sepúlveda joked to rescue workers. "You want to have the house all cleaned up before you leave."

Part of their perparations also included managing their new fame. While Sepúlveda had been the perfect host for underground videos and good for group morale, the men began to speculate that once above ground, they would need someone with a different skill set — more serious, more grounded in legalese. "I was able to talk to Mario about the group feeling that he was trying to take over the show. I told him, 'They are right. You need to back off. You are trying to take over. Maybe you don't realize it. You are always, always in front of the camera,'" said Ávalos. "It was an open secret that they wanted to beat the shit out of him."

On the last Friday, those concerns came to a head with a proposal to vote on a new spokesman. Was Mario really the right person for the new phase of media frenzy? Some of the men began to propose a more sedate, mature official voice. When that idea gained traction, a vote was called. Sepúlveda lost. The duty of official spokesman was handed to Juan Illanes, an erudite man filled with confidence, eloquence and a vague understanding of intellectual property and law. Sepúlveda took the decision like a slap to the face. To him it felt like a rejection of his leadership, an ungrateful snub. He immediately pulled back and began preparing for his journey into the media as a one-man show.

Day 65: Saturday, October 9

The rescue workers informed the men that the drill was now less than 33 feet from the tunnel. Sepúlveda sent a message up to the rescue team: "We are all going up the tunnel to watch the drill. When it comes through, we are going to dance and party all night. Tell them to stop sending down the palomas. No one will be there to receive them."

The food *paloma* was never left unattended. Even at the most stressful moments the men assured that their lifeline was under strict watch. Now they had decided to migrate several hundred feet higher in the tunnels, to the workshop, where the drill was about to break through.

The men gathered in anxious anticipation. So many things had gone wrong: broken drills, errant boreholes, the collapse itself. The men were incredulous. Was salvation really as close as it sounded? The men huddled in the tunnel, 160 feet from where they prayed the drill would rip through. A mixture of mud and water poured into the ramp.

Alex Vega started writing. He shielded a notebook from the dripping water and began to chronicle the historic moment, a minute-by-minute account addressed to his wife. A whirl of dust, the sharp noise of hammers pounding and flying debris filled the air. The miners all stared like children waiting for Santa Claus to come down the chimney. Their eyes were fixed as, drenched in sweat, helmets fastened and hands covered with thick work gloves, the men prepared for a final mission.

There was little they could do besides watch. Pedro Cortés spoke by telephone to the engineers above, providing minute-by-minute updates that were translated and then relayed to Jeff Hart, the driller, who adjusted the speed and pressure on the drill. The flying debris and raucous noise kept them from creeping too close.

The miners listened to the slower revolutions of the drill as it drew closer. And then the sound changed to the scream of metal fighting metal, an uneven screeching that set the men's teeth on edge. Jeff Hart had no option but to try to grind through the mess. Everyone was remembering the earlier incident when the hammer had sheared off, causing a four-day delay.

The drilling resumed and the drill again jammed. Again, that shearing sound. Metal roof bolts on the inner ceiling of the cavern were entangled with the hammerhead. Pedro Cortés stayed in phone contact with the engineers above. They told him the drill was less than a yard away.

Above ground, Jeff Hart slowed the drill. If he advanced too fast, the drill might pass straight through the roof and become jammed; wrenching the bit loose might shatter the fragile sections of the tunnel. The final inches were painstaking. With Sougarret, Golborne and a growing crowd of rescue workers and government officials clustered around, Hart stopped repeatedly to check the video shot from below. Like air traffic controllers guiding in a floundering flight, the miners worked and prayed for the safe arrival of the drill.

Underground, the sound was deafening. Even with earplugs and a second layer of protection in the form of earphones, the roar of the drill chewing and pounding the rock was painfully loud. This time the drill overpowered the bolts and at 8 AM, it broke through.

As the nub of the drill appeared through the roof of the workshop, a massive cloud of dust filled the caverns. Many of the men had a flashback to the first cave-in, as the dust cut off their vision. Now, the dust storm was a glorious sign of freedom. The miners below hugged and hooted.

Then the word came from below: the drill was through! Workers at Plan B were slow to grasp that the mission had been completed. Hart joked, "I thought it was my heart exploding." Golborne and Sougarret began hugging. Then a champagne cork popped. A truck horn wailed. And finally the Plan B platform was filled with hugging and jumping helmeted workers. The men embraced and locked arms around each other's shoulders. They danced in circles. Throughout the valley, a cacophony of horns, bells and yells filled the air. After two months, the rescue tunnel had reached the men.

Hart immediately began to pack up. His job was done. Now it was time to let the Chileans take over. Wandering through Camp Hope in his oil-stained work overalls, he gazed in wonder at the hole he had drilled to the trapped miners' remote refuge. Hart seemed baffled by his instant celebrity status. Women hugged him; reporters shoved and grappled to record his every word. Yet Hart was unable to explain his talent. With eyes that said, "You will never understand," he looked at the reporters and stated, "I am a driller. If you are not a driller you can't understand me. It is a vibration that comes up from the ground. I feel it in my feet, and then I know where the drill is." Had Hart's drill slipped off course by 20 inches, he would have missed the tunnel. He had thrown a bull's-eye. Like a long-distance sniper, he was perfect.

Hart described in detail how the final moments of the drilling operation had been a joint effort with the miners below feeding him live video footage. Asked what he would tell the miners, Hart laughed and said, "Two days ago we sent them a message: 'We will be there.' Now I would say, 'Follow us!'"

Camp Hope erupted in joy. Rescue workers in hard hats went from tent to tent hugging family members. Family members of the trapped men were now letting their hope loose. "I am sending him tranquillity and comfort. The worst is over," said Alonso Gallardo, thirty-four years old, a nephew of trapped miner Mario Gómez.

"We are going to have a huge party in the neighborhood," said Daniel Sanderson, twenty-seven, who slept for only one hour during the night as he awaited the fate of his two best friends, who were trapped. Sanderson, who also worked in the San José mine, said that despite the dangers and the extreme trauma of being trapped underground for weeks, his friends would continue to work as miners. "They already wrote me that they are going to look for new mining jobs. We are all miners."

"These are for everyone," said Juan Gonzalez, thirty-nine years old, as he unloaded forty crates of fresh avocados at his family's tent inside Camp Hope. "I just want to hug them," he said in reference to Renán and Florencio Ávalos, his two brothers who were both trapped. "I would tell them to stay calm, we are all waiting here."

"If it is Tuesday, Wednesday or Thursday it does not matter," said President Sebastián Piñera. "What matters is to rescue them alive and rescue them safely. And for that, we will spare no effort." What Piñera did not clarify was whether the rescue plan now included lining the entire shaft with metal tubing or just portions of the shaft. The tubing had long been considered an essential part of the rescue operation, a guarantee that the walls of the shaft would be smooth enough to provide an uninterrupted and easy course for the wheels of the Phoenix to follow. Now the tubing was being reevaluated. Engineers feared that the slight bends and twists in the shaft would complicate the installation of the tubing. What if a single segment buckled and became jammed? Was it riskier to line or not to line? That became the question.

Minister Golborne also preached caution. "This is an important achievement, but we still have not rescued anybody. This rescue won't be over until the last person leaves the mine."

Even as he spoke, family members gathered around the embers of campfires, ate breakfast with smiles and shared coffee and hugs with strangers. Hundreds of foreign journalists rushed to file the news that Los 33 had moved one step closer to freedom.

Below ground, at the bottom of the mine, Claudio Yañez photographed the rescue shaft, although it was impossible to see more than a few yards up before darkness swallowed all detail. Together with Samuel Ávalos, they made a home video, reaching with the camera into the shaft, as if by sheer effort alone and a leap of imagination they might be instantly delivered to that lost world above. The rescue shaft was 28 inches across, large enough to deliver a welcome breeze of cold air into the tunnel. The men marveled at the pleasure of semi-fresh, cool air. The rescue shaft was like a crude system of air-conditioning. The men could not know that the same rescue hole that was so close to delivering them to safety was also a death trap.

The cooler air filtering in shifted the temperature inside the fragile mine. Cold air caused the walls of the mountain to contract. This violent change in temperature, so pleasing to the men, had the effect of destabilizing the entire mine.

Day 66: Sunday, October 10

More than ever the men felt trapped. Time seemed to stop. With no sun, no dawn, no way to mark the time, they would regularly ask each other if it was morning yet.

Then, at 6 AM, the early-morning peace was shattered by first one huge rippling roar, then another and another. "Richard [Villarroel] kicked my feet and woke me up. He said the mountain was coming to get us," said Samuel Ávalos. "I thought we were doomed. This whole thing is coming down. If it does, we are gone. The whole mountain was so unstable. Anything could happen. It did not stop. *Pow! Pow! Pow! Pow, pow, pow!*

It kept exploding."

Luis Urzúa called up to Sougarret. "The mountain is cracking, making lots of noise," said Urzúa, who along with the rest of the men had become alarmed as dust and a strange wind blew through the tunnels. Sougarret tried to assure them that the cave-in had happened far above them, that they were in no direct danger.

When Samuel Ávalos heard the collapse of the mountain, he was convinced that this was the final act — that the entire entrapment had been an inevitable road toward death. Ávalos was certain that the mine would come crashing down, that the mine was a living being filled with vengeance and determination to trap men inside.

Omar Reygadas, a fifty-six-year-old miner and union leader, was sure that the cracking and explosion of rock were a message from above. To Reygadas's ears, the cacophony of cracks and blistering snaps as rock let loose were nothing more than God's voice. "I am Christian. I thought it was a warning, that God had done a miracle for us and we had to keep believing in him. To thank God for giving us life and thank him for letting us out. The mountain was exploding. That we had promises to keep and we swore to be better people, and I think that the mountain was reminding us to keep our word. Others were saying, 'The mine does not want us to leave. The mine wants to keep a single miner here.'"

Richard Villarroel remained calm. He lay in bed, gathering strength for the journey up, confident that nothing could stop them now. He was determined to see his wife give birth to their son Richard Jr. The due date was less than two weeks away. Having survived the collapse, the starvation, the heat and the humidity, Villarroel felt invincible. All the cracking and the groaning of the mountain was unable to shake his sense that destiny had brought him this far because he was meant to survive.

By noon, the cracking slowed and all but stopped. But even the silence was a frightening reminder that the mountain was pausing between movements.

Few of the men slept that night.

TWELVE
THE FINAL PREPARATIONS

Day 67: Monday, October 11

The miners began preparing for the arrival of the rescue capsule. Spotlights ringed the area, making this section of the workshop look like a miniature stage. The shaft itself was both unremarkable and miraculous. At first glance, the rescue shaft was nearly invisible, just another dark splotch amid the folds and uneven geometric cuts of the ceiling. But the men visited the hole like it was a sacred shrine, making daily, sometimes hourly, pilgrimages. This section of the tunnels was high above the men's living quarters and was usually empty. With the drilling complete, a silence filled the air, the only sound the constant splat of water drops falling to the floor.

The men talked nervously, smoked and tried to make time pass faster. They discussed their pact — a vow of silence. Each man had promised not to discuss the details of their life below. Each miner had sworn not to criticize any other miner. They were free to speak about their own experience, but not to share with the media any details about their at times fractious existence. That, the men had decided, would be saved for their collective movie.

Franklin Lobos had been at the forefront of these discussions. Lobos reminded the men to stay united. He sought to create a nonprofit foundation that would lionize the men's accomplishments, showcase their survival and be housed in a museum dedicated to the drama. All proceeds from the movie they dreamed of would be divided into thirty-three shares, assuring that each man would benefit from the media exploitation. Known later as the "pact of silence," the agreement was intended to protect their privacy and cover up embarrassing incidents; rumors abounded that the men had partaken in homosexual dalliances, soft drug use and the occasional fistfight. But the core of the pact was also financial. The men viewed their experience as a collective suffering that demanded that the spoils be divided equally. The pact would not survive even the first twenty-four hours.

As the critical last hours slowly ticked away, the miners pleaded for more and more cigarettes. "This is not a stop-smoking program," said Dr. Romagnoli, as he crammed packs of cigarettes into the paloma. Asked about the irony of a personal trainer rushing deliveries of cigarettes to his patients, Romagnoli insisted, "This is a rescue mission. . . . I don't have the heart to take away their cigarettes." The miners were nervous but in good spirits, asking Romagnoli to send down pisco, rum and mixers.

Special waterproof clothing was also sent to the miners. Custom-tailored for each man with fabric imported from Japan, the green jumpsuits were rolled up and shipped down the paloma along with clean socks, vitamins and a pair of black Oakley Radar–style sunglasses.

In a telling sign of the men's humility, they asked for shoe polish. The men had lived like animals for weeks. Bacteria and fungus had invaded their lives, colonizing their skin. Now, with the world watching, the men sought the most basic levels of human dignity — a fresh face, clean hair and polished shoes.

Though the men were expected to be extricated at night, the sunglasses would serve to shield their eyes from the glaring floodlights now surrounding the rescue area. Family members expressed fear about another type of spotlight: the press barrage. In a poll by the Chilean newspaper *La Tercera,* family members said their fears of media "overexposure" outweighed their concerns about the men's psychological and physical health.

At the paloma station on the hillside above the San José mine, Dr. Liliana Devia rehearsed the evacuation protocol by laying a sketch of the field hospital on a desk. She then moved around pieces of colored Legos as she described the medical plan, like an army general preparing the troops for battle.

"This is the first time in many weeks that the miners are going to be completely alone," said Dr. Mañalich, the talkative Chilean health minister who feared the miners were so nervous they would suffer panic attacks during the ascent.

After numerous days of practice inside the Phoenix, a group of rescue workers were convinced that the experimental capsule was safe, solid and, although a bit cramped, not particularly uncomfortable. The idea of watching a monotonous wall of rock slide by for fifteen minutes was enough to make even the most experienced sailor seasick. The men would be advised to close their eyes if necessary. Thanks to Dr. Romagnoli and a sophisticated wireless transmission of vital signs, should any of the miners go into a panic attack, the indications would pop up on the laptop and Pedro Gallo or a doctor would try to calm the miner down.

A host of requests from the miners for specialized soundtracks and songs for their ascent helped assuage fears that the men would be unable to handle the fifteen-minute solitary voyage. Victor Zamora pleaded to have Bob Marley's "Buffalo Soldier" blasting through the capsule as he surfaced.

If all went according to schedule, the men would be removed at the rate of one every ninety minutes, a roughly two-day marathon in which the already flagging endurance of the entire team would be tested.

The Copiapó hospital, where the men would be taken via helicopter, was bracing for a siege. Security barriers were erected and neighbors cashed in, renting out overgrown backyards to satellite crews and broadcasters around the world. Windows in two wards were taped over and heavy curtains installed to shield the men's sensitive eyes from both sunlight and the invasive probing of long-distance lenses.

Authorities pleaded with the media to give the men time alone with their loved ones, but given the appetite for the story and the intense competition for the first interviews, few reporters seemed willing to comply. In addition to stories alleging homosexual activity and drug use in the mine, the miners were expected to be grilled about their often complicated home lives above ground. The mélange of lovers, wives and a recently discovered love child made the arrival anything but relaxing. "I was waiting for them to ask us who wanted to go up," said Sepúlveda. "I think about ten of us would have chosen to stay underground."

In the center of Copiapó, less than a mile from the hospital, hundreds of miners protested and marched, snarling traffic. These men worked for San Esteban Primera, the holding company and owner of the San José mine and several other local mines and processing plants. While the press and lawsuits were focused on the thirty-three trapped workers, another 250 workers were out of work and out of the limelight. They were demanding that their salaries be paid and paperwork finished so they could look for new jobs.

Blowing horns and carrying signs reading "The 33 are fine, the others are screwed," the men tried to draw attention to the broader consequences of the San José collapse.

"They will get no international trips, no

presents, no TV invitations, exclusive interviews or special treatment," read an editorial in the local newspaper *El Atacameño*. "They and their families are just waiting to return to normal lives, to land dignified jobs that will let them get ahead."

Even as the unemployed miners marched in the streets, the wives of the trapped miners were being polished for the spotlight. Copiapó mayor Maglio Cicardini had organized free spa treatments for the wives. "I decided that they should have a beauty session," said the flamboyant mayor. As the women emerged from the salon, Cicardini said, "They all look so beautiful and striking that I doubt even their husbands will recognize them."

Up at the mine site, the entire mountain was on full alert. Hundreds of rescue workers were preparing for jobs large and small. The Chilean Air Force had helicopter pilots ready. In the field hospital, twenty-four doctors were on call. A platoon of nurses and paramedics were manning stations to measure blood pressure, deliver glucose and conduct overall physical exams of the men.

Six different command centers were fully staffed with personnel ranging from air traffic controllers to a team of surgeons. The Chilean investigative police (PDI) had a team ready to fingerprint and photograph the miners as soon as they were rescued. "The idea is to verify that the people who are inside the mine are, in effect, the same names that we have all assumed," said Óscar Miranda, a police inspector.

Police patrolled on horseback and motorcycle and on foot, scouring the hills for infiltrating journalists. Wireless transmissions used by the government were limited to key information only; for the past weeks the government had lived in fear that reporters would develop the ability to intercept wireless communications.

Dr. Romagnoli was watching a computer screen that printed out live updates on the men's vital signs. He could see the blood pressure and heartbeat of the men rising. He literally kept the pulse of the operation. Mario Gómez was suffering from shortness of breath, his silicosis exacerbated by the stress of imminent rescue. Sepúlveda had not taken his medication to keep his exuberance under control and was as hyper as ever. Osmán Araya was groaning from the pain of his infected tooth. All the miners were told to stop eating eight hours before the rescue commenced; like patients before surgery, they were expected to follow strict medical instructions.

Yonni Barrios was no longer on call. The stress of entrapment had finally ruined his ability to treat others. In fact, his life above ground was complicated by the much publicized battle between his wife and his lover, both of whom took turns trashing the other in the press or destroying the shrine and photographs left by the other. For Yonni, the situation was exhausting. He no longer had the strength to monitor the miners' health and distribute medications.

At 3 PM the men had one last task to complete before the rescue could advance: a final blast of dynamite.

The rescue capsule was so wide it couldn't descend low enough for the men to climb inside. Instead it got stuck on one of the walls. The miners were asked to place explosives to blow away a section of the solid rock wall. For the experienced *cargador de tiro* (master of explosives) the order was routine, hardly different from a mail room clerk being asked to dispatch a mountain of letters.

Miners trained in the use and transportation of explosives gently filled the paloma with blasting caps and enough explosives to remove the tons of rocks that were impeding the rescue capsule from entering the rock cave. The miners had detonated charges during their confinement — both to send SOS messages during the first chaotic hours of the entrapment and for more sophisticated engineering operations in the ensuing weeks. Publicly the Piñera government denied reports of these repeated detonations in an effort both to limit questioning of the rescue scenario and to calm the already frayed patience of desperate family members.

Once they had gathered and stashed the explosives below, the miners needed to drill holes to implant the dynamite in the rock walls. From the paloma that delivered air and water, Pablo Rojas received a tube that delivered compressed air and Victor Segovia connected the air to a compression drill. Segovia was surprised that the makeshift drill sliced easily into the solid rock. Segovia perforated and Rojas stuffed the six holes with sticks of dynamite. A single fuse connected the explosives.

With Urzúa and Florencio Ávalos supervising, the rest of the miners gathered in the safety refuge — standard procedure anytime they "burned." Rojas lit the fuse and they hustled to the safety refuge. Fifteen minutes later, a short crack signaled the explosion was over. The miners all rushed to see the results. When the dust cloud settled, they smiled. The explosives had blasted away a section of rock wall. Now when the capsule arrived, it would not be jammed against a wall.

The miners began to pile up the debris and wreckage to create the necessary, yard-high landing pad. The idea was to have the capsule lowered down so that it sat on the ground without the uppermost portion completely clearing the hole. The miners could simply open the door, be strapped in and instantly hauled out without worrying about the capsule swinging. Using heavy machinery, the miners piled up the debris and prepared the landing pad.

As the men excitedly assembled the pad, the mountain once again began moving. The dynamite had not only removed a portion of the wall, it had sent a short, sharp vibration through the tunnel. That vibration now loosened rocks, causing first a drizzle and then a roar of rock slides inside the tunnel. The lowest level, down by the pool, collapsed. Slabs of rock between the refuge and the workshop also gave way, sending a wall of rocks spilling into the main avenue. The mountain had started to cry.

The men put their helmets on. No one was sure if this was a brief sob or if the whole mountain would start to wail and bombard them with its deadly tears.

Luis Urzúa was planning his last day below. As shift foreman, he had been overshadowed in much of the day-to-day decision making. In terms of charisma, he could not even walk the same stage as Sepúlveda. Yet he still maintained a power and dignity based on the hierarchy of the mining culture, which demanded respect for the shift foreman. The men accepted that Urzúa would be the last man to leave the tunnel, like a ship's captain who first sees to the safety of his crew and then saves himself.

In a brief conversation with *The Guardian* newspaper that Monday, Urzúa gave his first interview since the ordeal began. "We had a stage of our lives which we never planned for and I hope to never live again . . . but that is the life of a miner," he said. Asked about the dangers of the San José mine, Urzúa said, "We always say that when you go into a mine, you greet the mine, ask permission to enter and respect the mine. With that, you hope to be allowed out."

Day 68: Tuesday, October 12

At 7 AM, the scene inside the refuge looked like a refugee shelter. Clothing was strewn about, and rows of men nervously tossed and turned on their cots. Nearly naked, wearing only shorts, the men stretched out and shielded their eyes from the permanent lighting inside the shelter. The cots were jammed together — if a man stretched his arms, he would touch companions on both sides.

After hours of nervous pacing and card playing, many of the men had finally collapsed in sleep. Claudio Bugueño and Pedro Cortés read the newspaper with their lamps, killing time and attempting to stave off nervous anticipation. Victor Zamora cracked jokes and explored the humid cave.

The habitual music had been turned off. The incessant drilling of the past months was finally gone. For the first time in their entire odyssey, silence was a welcome companion.

With the rescue scheduled to begin within twenty-four hours, the world was quivering with anticipation. At Camp Hope, the only street was cordoned off by barriers in a failed attempt to keep the area clear of journalists.

In addition to the ubiquitous regulars, the press corps was augmented by a bevy of beautiful TV hostesses. They posed like peacocks on the rocky outcroppings, reporting the story to millions of viewers worldwide. Where they came from remained a mystery. Had they parachuted in overnight?

For months the Camp Hope press crew had been a grungy, mostly male enclave. With showers scarce and dust abundant, the fashion styles ranged from combat chic to casual mountain climber. Now an entirely new breed had arrived, epitomized by NBC's Natalie Morales, who strutted around with model measurements, gleaming teeth and perfect hair.

An auction began in Camp Hope — which miner would be the first to sell his story to a tabloid? Rumors swirled that a German tabloid had offered $40,000 and that a trapped miner had already signed the contract. Family members began tempting the press with offers of exclusive photographs and footage shot below ground.

For weeks, Luis Urzúa had been complaining about the number of cameras circulating below. "My husband told me in a letter that everything was getting searched in the palomas and that I should be careful. So it was my idea to put the camera inside a pair of socks," one wife told Chilean newspaper *The Clinic*. "The photographs are going to be useful as evidence in case of any legal settlement for the entrapment. Now every time we send letters and we mention the camera, we speak in code. We call it 'the toy.'"

For Carolina Lobos, the press siege was

too much. She had fallen into the press net early in the drama, giving numerous interviews and even appearing on *Who Wants to Be a Millionaire.* She won $25,000 on the game show. Now she was fleeing the press.

"My father was famous as a football [soccer] player, but now he is a miner. He knows the double-edgedness of being in the news. He might be a hero but I don't want press. I just want to disappear," said Lobos, who was planning a low-key escape with her father and family. "He is very upset by the showbiz angle this has all taken. What he lived was traumatic, and all of this is distracting from the real mission — the rescue. . . . My dad never lost his bearings. He always realized it was an accident, not a show."

Though the rescue was set to begin in less than twelve hours, the Phoenix capsule now lay prone on the floor of a workshop high on the hill. Workers had disemboweled the electronic inner workings and were installing a camera on its roof, after realizing at the last minute that the capsule had no capability to film up the shaft. In the event that rocks started falling or the walls of the hole collapsed, it was critical to have the ability to monitor the situation.

Five technicians, led by Pedro Gallo, were fussing over the capsule's retractable wheels, the audio intercom and the new camera. The capsule looked more like a prototype than ever before. Would it be ready for the 11 PM rescue launch? No one dared ask.

In the early afternoon, the miners threw a monkey wrench into the rescue plans. Signs of rebellion echoed from below. Los 33 had decided to boycott the helicopter flight to Copiapó. They were arguing for another scenario: all thirty-three would assemble at the field hospital. No one was flying to safety until they had all been reunited at the rescue site. Further rumors swirled. The miners were demanding to walk down the hill together in triumph; having entered the mine together, they would leave the mine together.

Medical and psychological teams hurried to dissuade the miners. While the health of each miner was reasonably stable, too many uncertainties existed to allow the men to simply pop out of the ground after ten weeks and walk off toward the horizon. What if they had lingering health issues that had been misdiagnosed below or never diagnosed? Was it responsible to let the men fulfill this understandable but whimsical desire? The field hospital had been built for a capacity of sixteen men. Thirty-three would not fit.

The ACHS insurance company began to frantically consult lawyers. Could the miners be threatened with the suspension of health and workers' compensation insurance? The answer was no. Alejandro Pino, the ACHS lead logistical coordinator, began to gather a fleet of ambulances. In case the men successfully boycotted the helicopter flights, he wanted to have his own Plan B up and running.

Iturra, the psychologist, had his last placid conversation with the men. He spoke to shift leaders and encouraged them to keep the men busy. He suggested the men take a nap. It was the final time they would ignore his advice.

Once the last adjustments to the Phoenix were completed, the capsule was loaded with 176 pounds of sand and then lowered all the way up and down the shaft. Down in ten minutes. Up in ten. The Phoenix operation was so smooth that instead of forty-eight hours, it now seemed possible to complete the entire rescue in half that time.

At 7 PM, Chilean rescue leader André Sougarret sent a Twitter message that the men "have spent their last night underground." Chilean president Sebastián Piñera could hardly contain his enthusiasm as he announced the rescue of the thirty-three trapped miners. Nearly two months earlier a miner had pleaded with President Piñera over a makeshift phone line, "Get us out of this hell." Now the president had ridden the drama to worldwide fame and higher poll ratings.

The worldwide audience began a collective countdown to a never before seen rescue attempt — lowering the Phoenix nearly a half mile deep, strapping the men in one by one, then using a high-tech Austrian winch — with the finest German cable — to hoist the men to freedom.

After nearly ten weeks at the bottom of a collapsed copper and gold mine, the men were now awaiting a single, final challenge — to board the bullet-shaped rescue capsule and careen up a series of curves and slants to escape from the underground prison.

"There are people in favor of having Mario Sepúlveda be first," the psychologist Iturra had told the press days earlier. "They suggest that Mario Sepúlveda narrate the ascent of each of his companions, or at least some of them. But I told Mario to remember that he is going to arrive very tired, and if he comes out too much in the press, the price he can charge as a celebrity will go down."

The Chilean government decided to make Mario the second to arrive. He was clearly the most famous, but just in case any problems arose, the excitable Sepúlveda was not the best candidate to go first. Instead, assistant foreman Florencio Ávalos, who exhibited a rare combination of street intelligence, physical endurance and mining experience, would be the first of the thirty-three trapped miners to be hauled up.

Rescue workers, led by hyperkinetic President Piñera, chose Ávalos because if anything went wrong, Ávalos was expected to maintain his cool and feed information to the control center organizing the hundreds of men and women involved in this complex rescue operation.

Political considerations also led the Chileans to put Bolivian miner Carlos Mamani in the first group. "We can't put him first because then they would accuse us of using 'the Bolivian' as a guinea pig. And if he went too late, then we would be called racist, so the government has decided he will go in the first five," said a doctor on the rescue team who asked not to be named.

Evo Morales, the Bolivian president, had asked to be present to receive Mamani, and the Chileans eagerly agreed. With a century-old dispute over Bolivian access to the Pacific Ocean now at a critical negotiation stage, any opportunity to foster mutual understanding was cherished. Piñera warmly welcomed President Morales, to the chagrin of Mamani, who despised the political scene and Morales in particular.

When Dr. Mañalich spoke to the miners, several of the men expressed the desire to be the last man out in what he called "a completely admirable show of solidarity." On further questioning, however, the men's real motivation was revealed: a guaranteed place in Guinness World Records for the longest time a miner had ever been trapped underground. Given the complexities of the current situation, it would be a record that many expected to be unbreakable. The situation was resolved when Guinness accorded the record to the group as a whole, not to one individual miner.

At 8 PM, a group of five rescue workers crowded into a small white container on the hill. The men gossiped about the imminent journey and chatted about how a trial run had left one rescue worker so dizzy he vomited inside the capsule. "It is far wetter than you'd think," said one uniformed man, describing a trial run halfway down the shaft.

"My clothes were soaked."

"We have divided the men into two groups: GOLF — the healthy miners — and FOXTROT — the potential problems," explained Dr. Liliana Devia, as she began the final briefing on the health and welfare of the miners. Dr. Devia warned the rescuers that several of the men were in far worse condition than the press or even their families realized. One of the men was described as bipolar. Another was said to have attempted suicide years before. A third had told the psychologists and nurses that he had seven lovers awaiting him.

As she described the nine unhealthy miners in the FOXTROT group, Dr. Devia noted that two miners would immediately be sent for dental surgery. Several others were so nervous and fragile there was concern that they might turn aggressive. Sedation medicines were primed. "The needle is ready," said Dr. Devia, who briefed the rescuers on how to administer drugs that "will leave them ironed flat to the bed."

Devia, who was in charge of briefing the rescue workers on the latest health and mental conditions of each miner, outlined the rescue protocol. The first rescuer was designated as both doctor and policeman. He was to monitor the miners' health and keep them in line. Multiple video cameras at the bottom of the mine would provide a live feed, allowing psychologists, doctors and mining engineers to monitor the operation in real time.

Should tensions erupt or an accident shatter protocol, the rescuers were authorized to maintain order, even to sedate the miners. If nothing went wrong, rescuers would brief the miners about the operation of the Phoenix and instruct the men to strap a girdle to their stomach and to put on long stretch socks that came up to their upper thighs. Prolonged standing was not a problem for the men, but the panty hose–like socks — were to help with blood circulation. The girdle was instrumental in shrinking the diameter of the men enough so they would fit into the tight confines of the Phoenix. Once they were strapped in, the door would be latched shut and a signal given, and the men would embark on their journey to freedom. For fifteen minutes they would swoosh and bank as the capsule careened skyward.

Upon arrival topside, each miner would be immediately helped from the capsule and met by President Piñera, then taken to greet their families for a brief hug and kiss. Then they would be placed on a stretcher and wheeled into a fully staffed triage field hospital organized by the Asociación Chilena de Seguridad (ACHS), which had been built just 65 feet from the rescue hole. The men would be wheeled into the hospital, briefly examined and then given a simple yet profound pleasure: their first decent shower in ten weeks.

In case the men were harboring more serious ailments — either physical or mental — the medical team would keep them several hours for observation, then send them to a second set of modular buildings (also assembled in a record time of less than a week) where they could have a longer visit with family. The final stage was a short drive to the highest point of the rescue operation where a helicopter landing pad had been built. Instead of the hour-long land journey, which was bound to be flooded with paparazzi, the men would board helicopters from the Chilean Air Force and be shuttled to an army regiment near the public hospital in Copiapó. Again the men would be interned, this time for more serious blood tests, lab work and longer sessions with psychologists.

Throughout the planning of the rescue, the health officials were aware that they had no legal power to force the men to accept medical help. If a miner became feisty and demanded to exit the Phoenix and walk home, the authorities had no legal grounds to prevent him. Psychologically, however, the men were expected to be not only grateful but dependent to the point of obsequiousness. Having been saved from certain death, the amount of gratitude that continually flowed from below was sufficient to imagine there would be no difficulty in leading the men through the now extensively rehearsed protocol.

Down below, the men were now in a party mood. Music blasted from the mini speakers set up by the rescue spot. Video cameras filmed every last detail. The men posed for final pictures, and a sense of nervous anticipation filled the humid air.

The miners gathered to hear the final review of the escape protocol. The capsule had been designed with a floor that opened up to allow the miners to drop out the bottom. In the worst-case scenario, if the Phoenix got stuck, the occupant was expected to lower himself back down to the bottom of the mine while engineers reconfigured the capsule.

At 8 PM, President Piñera set up shop at a tent halfway up the hill. Like a venue for a low-budget wedding, the mountainside had been decked out with portable tables, long blue tablecloths, soft drinks, juices and snacks and a pair of flat-screen TVs. Here the family members would spend their last agonizing hours watching and waiting. Piñera, his wife, Cecilia Morel, and top aides were set to greet the families here and, if necessary, monitor developments from a live video feed that provided shots of the underground scene.

Down at Camp Hope, the families were trapped. Each family was ringed by a cluster of journalists who clamored for a tight reaction shot or a final comment about the stress of waiting sixty-nine days to see a loved one. At the Ávalos family tent more than a hundred journalists balanced precariously on stepladders. Shoving for better access resulted in a journalist toppling the entire tent, breaking eggs, upending makeshift shelves with food and nearly crushing the Ávalos family in the process.

No one was too upset. The families and press had learned to live with and understand each other, despite language and cultural barriers. But not all guests at Camp Hope were welcome. When President Piñera's brother, Miguel Piñera, known as "El Negro" ("the Black Man"), arrived, some family members erupted in protest. The infamous sibling had picked up the nickname for either his jet-black hair or his role as the family black sheep (depending on whose version you believe). Famous as a nightclub owner, singer and lover of all-night parties, El Negro was insulted and practically chased out of the camp. "Get out of here!" shouted one family member. "We don't want any more showbiz here."

A pair of Chilean Air Force helicopters swept in and out of the helipad, conducting last-minute practice runs to the hospital in Copiapó. By land the journey was a twisted, dangerous sixty-minute drive. By helicopter, the men would arrive in the emergency room within five minutes.

Then the mountain rebelled again. From below came word of another avalanche. The roof of the mine was again cracking and groaning. With a sound like the rumble of an avalanche and the crack of rocks crashing down, the blast of falling rock and the creaking of an entire mountain were reminders that salvation was still not guaranteed.

The Chilean government sought to censor the news of the mine collapsing at the last minute. Not now. Not when they were so close. But attempts at secrecy were futile because by now dozens of family members had inside sources in the rescue operation.

The rumors multiplied with viral fervor. Despite repeated attempts to control the flow of information, the fledgling government of Piñera was no longer able to control the spigot of information now flowing from both the mouth of the mine and the miners themselves.

Far atop the mountain, at the *paloma* station, rescue workers gathered in disbelief. The men were set to be saved in a few hours and now the gods were raging? Dark superstitions haunted the men, who had no doubt that their tormentor was an enraged female goddess, a wily bitch who ruled this — and all — Chilean mines.

For older, more experienced miners, the final round was a classic bout of miner mythology. The mine was often thought to charge a tax, a price of admission to those who dared to enter. Now the unspoken but widely shared fear was that the tax would be paid in the form of a human life, that the mine would never allow all thirty-three men to escape unscathed.

As the mountain continuously cracked, rescue workers scrambled to speed up the rescue plan designed with both the precision of heart surgery and the blind guesswork of a never before attempted operation. The miners were on the cusp of freedom, but the constant groans and creaking inside the mine were a terrifying reminder that time was running out.

The miners were hardly shaken by the latest cracks. By now they had grown accustomed to the rain of rocks. If it was not directly in their area of operation, the men felt safe — lightning bolts might be striking but until someone was hit, the miners felt they had dodged death. Psychologists routinely see similar behavior in soldiers at war who, after multiple exposures to live combat, are able to walk steadily while bullets whistle nearby.

Televisión Nacional de Chile (TVN), Chilean national television, was set to broadcast the entire operation live. TVN wired seven cameras, each with a unique perspective. The live shots would be patched together so that family members and the world could follow every detail of the rescue. Like the Super Bowl or the World Cup, no angle would be missed.

Fearful that miners might arrive unconscious or covered in vomit, the Chilean government maintained full control over the images the world would see. Health officials successfully lobbied against putting the men on the world stage until their initial condition was known. A huge Chilean flag was raised to block the view of the non-official press, a move that provoked chants and whistles of protest from the gathered press.

At 11 PM, as the press hooted and complained that they couldn't film a thing, TVN broadcast the winch lifting the Phoenix in preparation for its maiden descent. Despite heavy media exposure over the past week, the capsule maintained an air of mystery. It looked like a rocket designed by clever sixteen-year-olds. With fins at the tail and retractable wheels on the sides, the cylinder was designed to slide smoothly over the curves that contoured the 2,000-foot tunnel into a snaking trip. Coming up the tube would be like a crude amusement park ride.

President Piñera looked at the rescue capsule in earnest. He asked Sougarret if the capsule was really 100 percent safe. Sougarret assured the anxious leader there was no worry, little risk. Piñera repeated his question, insistent in his queries. "I wanted to go down," said Piñera, who admitted that he was enthralled by the idea of personally vouching for the safety of the Phoenix. Security aides to the president were apoplectic. Having already suffered in their attempt to protect a president who insisted on flying his own helicopter and scuba diving, they knew he was serious. So did Cecilia Morel, the first lady. She immediately picked up the scent of a risky folly. Catching her husband's eye, she told him to abandon the plan. "Don't even think about it," she ordered. Though it ran against his instincts, Piñera obeyed.

With a tinge of jealousy, Piñera watched as rescue worker Manuel González climbed into the capsule — the first man to attempt a complete journey from the rocky mountainside down to the unknown world where thirty-three men had lived in physical isolation from the world for sixty-nine days. A large yellow wheel above the capsule began to spool out the cable, which slowly unwound. The fins of the capsule entered the shaft, and the Phoenix dipped out of sight, with the whole world watching.

THIRTEEN
THE RESCUE

Day 68: Tuesday, October 12

As the Phoenix descended, three separate video monitors were being scrutinized above ground. President Piñera and his wife, Cecilia, were surrounded by top aides who peered at the live feed from the bottom of the mine. When the capsule arrived, all the action would be broadcast to the world. Piñera had overruled aides who wanted the live coverage to be limited to long-distance panning shots that offered no sense of emotion or drama. Immediately understanding the worldwide interest and inherent drama in the entire operation, Piñera successfully argued that it was a moment for Chile to showcase its "know-how." Not coincidentally, that was the same message that the fledgling president had sought to sell to the Chilean public as his strongest virtue. Not a politician known for emotional or tender connections to his public, Piñera's strength and political capital were largely encapsulated by this entrepreneurial spirit of "let's get it done."

The second camera was manned by Otto, a serious yet genial Austrian who was in charge of lowering and raising the Phoenix via a 2,300-foot cable. Atop the platform of his truck-sized control center, Otto placed a laptop with a live feed from below. Here he could not only receive audio feeds from below but also watch as the Phoenix arrived. The grainy black-and-white image looked to Otto like a remote-control vehicle that was traveling to another planet.

The final video feed was manned by Pedro Gallo, the humble inventor who had catapulted from hapless telecom guy to the top ranks of the rescue operation and into the hearts of the miners. Few of the rescue team workers had logged as many hours as Gallo in daily conversations with the miners. As a working-class entrepreneur, he was able to understand their gripes, relay their concerns and assuage their secret desires. Though Gallo would later deny it, the miners swore that it was Gallo who stuffed chocolates and sweets into the paloma. Those symbolic acts of defiance as well as Gallo's undisputed loyalty to the miners, and not the rescue hierarchy, had made him — in the eyes of the miners — a virtual saint.

Five rescue workers were now gathered, ready to head below. Two navy marines with extensive medical experience, two rescuers from Codelco and one member of GOPE, the Chilean police special operations unit that had bravely entered the mine during the dangerous first forty-eight hours after the collapse.

The Phoenix would arrive below through the roof of a workshop. When the mine was still functioning, the workshop had been a place to fix or store vehicles. During the men's entrapment the workshop was considered an area too unstable for sleeping, so the miners rarely ventured the 415 yards from their main living quarters up the tunnels to the workshop site. Now this dangerous area was ground zero for the final and most important day of the entire ten-week nightmare, and the men moved their cots and clothing to the area next to the workshop.

Despite the anticipation and adrenaline, regular shifts were maintained. Someone had to man the *paloma* to receive last-minute supplies — including special clothing, sunglasses and fresh socks. While food delivery would be suspended at the very last moment, the rescuers were expected to spend a full day underground and the paloma would be used to send hot meals to keep them fed and alert. The *paloma* shifts had been set in stone weeks earlier, long before the exact rescue date was set. For this last paloma shift it was Franklin Lobos who drew duty. It was an assignment that would nearly cost him his life.

At 11:37 PM a rattling and clanging alerted the gathered men to the Phoenix's approach. The red fins of the capsule descended from the roof as if in slow motion. As the capsule emerged bit by bit, it was like a visitor from another planet. The trapped miners were stunned. A dream come true. Yonni Barrios approached and peered inside at rescue specialist Manuel González. For the first time in sixty-nine days, another human had arrived.

The thirty-three watched in awe and respect as González unlatched the capsule's door, stepped down and hugged Barrios. A horde of nearly naked miners then rushed to hug and greet him.

For one miner, Florencio Ávalos, freedom was just minutes away.

Ávalos was ready. He had slipped into the tailored green jumpsuit with his name stitched across the chest. A pair of Oakley sunglasses protected his eyes. On his right wrist a monitor measured his pulse and sent wireless updates to the rescue team on the surface. His left index finger was inserted into a device that measured oxygen levels in his blood. Tightly wrapped around his chest, a sophisticated electronic monitor transmitted another half dozen vital signs to the technicians and doctors above ground.

The other miners gathered around to watch, photograph and make home videos of the scene. Despite their nervousness, a strange calm filled the chamber. Like professional athletes in a locker room before a big game, the men joked and paced but their confidence was evident. The men momentarily forgot the terror of the collapse and the lingering sensation that death had been stalking them. For now, the scene was more like a party as cumbia music blared from farther down the mine. White balloons bounced around the floor as the men ambled excitedly — naked except for pairs of clean white pants.

The prospect of escape filled them with a dose of adrenaline. The men now felt like they were going to actually win their ten-week battle with the mountain. Along the length of the dark tunnels, the miners made last-minute explorations of the tunnels, the bright beams from their flashlights dancing in the distance. The clanking of carabiners was a reminder that rescue workers from Codelco, GOPE and the Chilean Navy had arrived.

González placed a white plastic credential — like those used backstage at rock concerts — over the neck of Ávalos. The rescue was filled with formality, orders and procedures. Every detail had been rehearsed for weeks. Yet the mountain could still throw a monkey wrench into protocol. Even the deepest calm at 2,300 feet was a superficial escape from the claustrophobic reality.

At 11:53 PM Ávalos stepped into the capsule, and the rescue workers latched the door shut. The miners all listened impatiently to the chatter between Otto, the Austrian winch operator, the communications center and Pedro Cortés, below. Meanwhile Ávalos nervously anticipated the imminent family reunion: the two sons who had not seen their father for two months; the wife who had been writing letters and watching videos but had not touched or looked into the eyes of her husband. Ávalos had left for work on a cold winter morning; now it was spring.

As the capsule slid upward, Ávalos's *compañeros* screamed, cheered and whistled. Then, instantly, he was alone. For fifteen minutes, Ávalos peered through a metal mesh that sliced the world into diamond-shaped viewing holes. A light inside the capsule illuminated the smooth, wet rock walls. The spring-loaded metal wheels clanked as they rolled along the rocky path. The capsule dipped and bobbed as it followed the uneven tunnel and slowly brought Ávalos toward freedom.

When he was just 65 feet from the surface, Ávalos could see the first signs of light and hear the first sounds of life. Rescue workers were now screaming down, asking if he was okay. Then suddenly he was in the light: a hero to the waiting world, a father reunited with his crying sons and a huge boost in the polls to President Piñera, who waited in the front row.

As Florencio was pulled from the capsule, his nine-year-old son, Byron, broke down in tears. Rescue workers jumped and celebrated. The cameras flashed on a wrenching scene — for a moment the nine-year-old boy was alone, awash in emotions. First lady Cecilia Morel, health minister Mañalich and Rene Aguilar, the second in command of the rescue operation, swept in to calm the child. Then true comfort arrived — a hug from his father.

Ministers, hard hat rescue workers, doctors and journalists all openly wept at the beauty of the scene. The men had defined themselves from that first note as Los 33 and had been adopted by the world as a beloved collective, now famed for their ability to work as a team. In a world so often defined by bloody acts and individual egos, Los 33 remained united while entombed, a brotherhood of working-class heroes. Teamwork had kept them alive, and now they would all be rescued together.

Florencio hugged first his family, then President Piñera, then the rescue workers. Next he was placed on a stretcher and wheeled into the field hospital. The entire hospital staff erupted in applause. They assumed Ávalos was healthy — he had been chosen to journey first based on his mental and physical strength — but nonetheless he was given glucose and a nurse took his blood pressure. As he lay in the bed, Florencio thought about his younger brother Renán, still trapped below.

Thursday, October 14, 1 AM

As narrator, clown and undisputed leader of Los 33, Mario Sepúlveda had carried a constant burden on his shoulders for sixty-nine days. He had never failed to see the power of humor to guide the group — a court jester to the invisible kings and princes who sent orders from above. Yet Sepúlveda also had an instinctive ability, a native sense of group dynamics, knowing when it was necessary to use brute threats of physical violence. With the responsibilities of leadership lifted, he was about to bloom in the limelight.

Below ground, Sepúlveda had cracked a last few jokes before climbing into the capsule. Now, at 1:09 AM, as the Phoenix neared the surface, he began a running commentary on his own rescue.

"Hey, old woman!" he yelled to Katty, his thirty-three-year-old wife. Through the mesh, Sepúlveda could be heard laughing. As raucous cheers went up, Sepúlveda sprang from the capsule and, without pausing to let rescue workers remove his harness and safety jacket, bounded over to President Piñera, dropped to one knee and began pulling gifts from his cherished homemade yellow satchel: a handful of white rocks, gleaming with the golden sparkle of pyrite. A rock for the president. A rock for the minister. The recipients laughed and clutched the stones. Sepúlveda hugged a stunned Piñera three times, and then flirted with his own wife, suggesting they would have sex for so long that neither would be able to walk. "Get the wheelchair ready," he joked.

He danced over to hug Pedro Gallo, wrapping his arms tight and holding Gallo in deep appreciation for everything he had personally done to save the miners. Gallo wept. Sepúlveda led the crowd in a rousing cheer, a celebration of what a reporter at *The Guardian* called "a flash of global joy."

At Camp Hope, the delirium was brief. While the family members celebrated the first two rescues, nothing could be truly enjoyed until all the men were out. The precarious cable that separated life and death was still visible to all.

As the Phoenix capsule dipped back down into the mine, Ávalos and Sepúlveda were transferred from the triage unit at the field hospital to a welcome lounge higher up the hill, near the helicopter pad. Decorated with modern white couches, flower arrangements and ultra cool blue lighting, the ambience was akin to a chic after-hours club. There was no sense or smell of medicine, sickness or trauma; instead the Chilean mental health specialists had designed a gracious main reception area and then a wide hallway that led to private spaces.

In the welcome lounge, Ávalos huddled with his two sons, his wife and President Piñera. Across the hall Sepúlveda was in a similar family mode: laughing, hugging and kissing. Then Piñera pulled Sepúlveda aside and asked him to do a brief interview with a TV crew waiting in the wings. Without much option but to obey the president, Sepúlveda sat in front of the camera and described the experience as positive. "I am very content this happened to me because it was the moment in which I needed to change my life. I was with God and the Devil and they fought over me. God won. I took the best hand, the hand of God, and never did I doubt that God would get me out of the mine. I always knew."

Then Sepúlveda raced out to embrace Ávalos. The two men hugged as smiles filled their faces. The threat of the thirty-three men to stay united on the hilltop until all had arrived above ground seemed forgotten. The preparations by rescue workers — consultations with lawyers, threats to suspend health care, the convoy of ambulances at the ready to deliver the thirty-three by land — were all unnecessary. Sepúlveda and Ávalos strode proudly toward the helicopter. The rush of emotions and the gratitude they felt in the moment had erased all presumptions of a miner rebellion.

Juan Illanes
Carlos Mamani
Jimmy Sánchez

Osmán Araya
José Ojeda
Claudio Yañez
Mario Gómez
Alex Vega
Jorge Galleguillos
Edison Peña
Carlos Barrios
Victor Zamora
Victor Segovia
Daniel Herrera

One by one, the men were rescued with military precision. Each man had his story, his family and the emotional first hug, first kiss. Some dropped to their knees and prayed, others cried. It was enough raw emotion to make the world stop and watch in wonder. Hour after hour, the world was captivated by a shared sensation of compassion.

The Phoenix, its Chilean flag motif ever more battered and scratched, was a modern workhorse: firm, unfailing and loyal.

Miner after miner climbed into the capsule and rode to freedom. The men had doused themselves in a cheap cologne that had been smuggled down to them. They were not economical in their pursuit of a sweet smell. "God, the capsule stunk of cologne," said one of the rescuers. "Whatever it was, they were all using the same brand. It was overwhelming."

Richard Villarroel, took a final series of photographs before he left. He wanted to capture the last images of the refuge, his bed, his friends hugging and smiling and posing. The men had decorated the safety refuge like a museum exhibit, the walls hung with the flags of their favorite soccer teams as well as huge thank-you notes to the rescue team.

Entering the capsule with headphones strapped to his ears, Guatemalan crooner Ricardo Arjona singing in his head, Villarroel said he felt a sadness. "It was painful to see my friends below as I was leaving them." But as the capsule drew toward the surface, Villarroel began to scream with joy. He cursed the mine. "Then I felt a change in the air. Fresh air — that was my favorite moment. . . . What a difference."

For a worldwide audience estimated at one billion viewers, the Chilean mine rescue was picture perfect. The grainy video footage from underground seemed like a live shot from another planet. To many viewers, the drama and collective excitement was reminiscent of the first Apollo landing in 1969, when Neil Armstrong took those famous first steps on the surface of the moon.

At the bottom of the mine, however, the script was unraveling.

At 1:30 AM, as the capsule came down to pick up Omar Reygadas, miner number 17, a sharp crack echoed through the tunnel. Then came the crash of boulders and the rumble of an avalanche. The camera filming the rescue went blank. Now Operation San Lorenzo was blind.

At the head of the telecommunications post, Pedro Gallo immediately called by intercom to the miners below. He asked Pedro Cortés, who had helped wire the underground telecommunications hookup, to investigate. Cortés was hesitant; the fiber-optic cable was close to the recent avalanche. The dust had not even settled and now he was being asked to enter a potentially deadly zone of the tunnel.

"You're sending me down there? You know it's not safe in there. There've been two avalanches," Cortés stammered. Earlier in the year he had lost a finger inside the mine; now he was being asked to risk far more.

Gallo told him that a live video feed was crucial. The winch operators needed to see the operation live so they could gently guide the capsule to the ground. A rough landing could damage or jam the Phoenix. President Piñera and approximately one out of every four adults on the planet were watching.

Cortés reluctantly agreed to run the ultimate obstacle course. He would have to negotiate a gauntlet of falling rock from the recently collapsed roof, cracked and still cracking walls and then traverse a muddy 200-yard stretch. As he followed the fiber-optic cable, Cortés found the problem: a rock fall had sliced the cable.

There was no possibility of repairing the damage. Hundreds of pounds of rocks had buried and destroyed the line. Gallo thought for a moment and then figured out an instant solution: he could take a cable that fed the camera in the safety refuge some 1,000 feet below, disconnect it from that camera and have the miners rewire that same live fiber-optic cable to the main camera filming the rescue.

Gallo called down to the phone in the refuge and was shocked when Franklin Lobos picked up. Lobos was alone at the far end of the tunnel that had already suffered two avalanches. "Franklin! What are you doing there?"

"It's my shift. I'm receiving the food for the rescue workers," said Lobos, loyal and stoic. "Duty is duty and it is my turn. I have to complete the shift."

"Old man, you're nuts! There have been two collapses! Get out of there. Now," Gallo screamed into the phone.

"But the food? What about the food for the rescuers?" Lobos was stuck on protocol, unconcerned or unaware of the looming danger.

"Forget about it," Gallo yelled. "I'll send food down in the capsule. Get out!"

As Gallo scrambled to configure a new fiber-optic system, President Piñera, Televisión Nacional de Chile and Otto, the winch operator, all had the same urgent question: What happened to the image?

Gallo told Piñera and Otto the truth — that they had lost the signal and were working to reestablish a live image of the miners below. To TVN he simply reloaded a video clip from earlier in the rescue. "They were going crazy that there was no image, so I took some earlier shots and put those on the air. Then I asked them if they had image and they said thanks." A billion viewers around the world were also tricked. They never realized that the image of perfection being broadcast was a rerun to cover up a dramatic chapter far too risky for the Chilean government to allow the world to see. Like all reality television, the miner drama also required sleight of hand, editing and a script.

But there was no putting the actual rescue on hold. Beginning with Omar Reygadas, three miners were raised to safety without the benefit of live video from below.

"My ride up was pretty anxious," said Omar Reygadas, the seventeenth man to be rescued. As Reygadas was preparing to enter the Phoenix, the door jammed. Rescue workers could not get it open. Using a crowbar, they pried the metal mesh door open. "I thought the mine did not want me to leave," said Reygadas. "After they pried it open, they could not get it shut, so they used a plastic strap. I held the door as I went up so it would not open."

As the capsule rose, Reygadas began to taunt and joke with his companions below. "I was yelling to the guys below things like 'Fuck, I am out of here. I made it. I made it! I made it.'" Despite the overwhelming joy, Reygadas also felt an instant nostalgia for his underground world. "We were leaving something behind. We had lived there for a long time. I had a feeling that I was leaving part of myself inside there. It was sixty-nine days. Part of me stayed down there. I told myself that it would be my bad characteristics and that I would arrive at the surface with all my best characteristics."

Reygadas, a widower, was eager to hug and greet what he calls his "little monkeys," a troop of grandchildren. As he neared the surface, Reygadas began to scream to the rescue workers above. He would yell, "Chi . . . Chi . . . Chi . . ." and the responding "Le . . . Le . . . Le" was confirmation that he was nearly safe. "I heard a voice from above asking if I was okay, and I screamed, 'Fuck yes,' then I remembered the president was up there. . . ."

While Reygadas celebrated with his "little monkeys," Pedro Gallo had to ask Cortés below to attempt another suicidal mission. This time, instead of skirting death halfway down the tunnel, to the fiber-optic station, Gallo asked him to run the entire 400 yards to the safety refuge, disconnect the cable and rewire the camera.

"Don't send me again," Cortés pleaded. Then he agreed to again run the gauntlet. But first he wanted to say goodbye. Cortés put his face close to a second camera that was operating underground and said, "If something happens to me, here I am for the last time."

Gallo shivered with fear. He had sent Cortés on the mission; now he felt the weight of fate. If Cortés was crushed, maimed or killed, it would be on his conscience.

The avalanche had not sealed the tunnel — a fact confirmed when an exhausted Franklin Lobos arrived from below. He told Cortés that there was enough room to get by the two rock slides, wished him luck and prepared for his own escape. Cortés did not question the order; instead he prayed for his life, and prepared for one last trip to the refuge.

In a mine known to attract kamikaze miners, Cortés would now defy the gods a second time. Even under normal conditions the mine was capable of killing and maiming. Now, in this final act, it was ever more unstable and dangerous. Cortés survived the nearly hour-long journey and returned to a hero's welcome.

"I had his life in my hands," admitted Gallo, who at that point had been awake for more than forty-eight hours straight. "But it was a duty, and he had to carry it out."

With the cable in hand, Cortés and Ticona connected the camera. Then Gallo reminded them that TVN was broadcasting a "live shot" that showed an empty screen — no capsule, no people. In reality, a number of miners and rescuers were waiting and milling about. If Gallo suddenly flipped to the real action, he would blow the charade as figures suddenly popped up on a few hundred million TV screens.

The stage was cleared, the live shot hooked back in, and then miners and rescuers were allowed to wander back into the shot. "They never noticed," said Gallo with pride.

With increasing concern about the stability of the mountain, the rescue was speeded up. Instead of a leisurely pace, Operation San Lorenzo took on a new urgency. Bringing the first sixteen men to the surface had been a showcase to the world of Chilean efficiency and international cooperation. Now the vengeful mountain was threatening to drag the worldwide audience into a *Titanic*-sized tragedy. If a landslide were to smother the men at this last moment, it would also bury and kill a rare moment of global optimism. The atmosphere inside the mine still appeared cheery — music and balloons still bounced off the walls, but the sensation that the irate mine had one last round of surprises for the men was pervasive.

Esteban Rojas
Pablo Rojas
Darío Segovia
Yonni Barrios
Samuel Ávalos
Carlos Bugueño

José Henríquez
Renán Ávalos
Claudio Acuña

The rescue plan had been designed to include rescue workers trained both in the techniques of climbing and also in battlefield medicine. The Chilean Navy had sent two Marine Special Forces commandos with extensive medical background; they could handle any medical emergency and were loaded with everything from a locked box containing morphine to a needle loaded with anti-anxiety drugs. But in deference to local sentiment, Minister Golborne at the last moment broke with protocol and instead allowed Pedro Rivero, a local rescue worker, to rush to the bottom of the mine to help out. Rivero had risked his life in early attempts to find the miners and was a representative of the regional rescue corps. No one could question his bravery or his technical rescue skills. His timing, however, could not have been worse. With military-like precision, the entire rescue protocol had long been decided; now, Rivero's improvisational appearance sent a sliver of chaos into the finely tuned procedures.

Rivero stepped from the capsule and immediately caused problems. He brandished a camera, started filming and headed into the depths of the mine, the same tunnel that had just collapsed twice. Rivero's mission was, according to Pedro Gallo, who watched the whole scene, to film the last scenes in the refuge. None of the miners or rescue workers thought this was sensible. "Never rescue a rescue worker" was a motto for the entire team. With avalanches already threatening the integrity of the operation, an added risk like that taken by Rivero was seen as mad.

When Rivero returned, he asked for the phone and declared that he had been sent on a special mission by Golborne himself; now it would be his job to stay below until the end. According to Rivero, he would be the last man out. The navy men were dumbstruck. From a military point of view, Rivero's actions were close to treason.

A raging argument broke out. The navy men threatened to stuff Rivero into the capsule by force.

As he coordinated phone calls with Pedro Gallo, Cortés heard the raging argument nearby and was stunned by its source.

"What's happening?" Cortés asked Gallo. "The rescuers are arguing; didn't they come here to rescue us?" The miners gathered to watch the bizarre spectacle.

A call from Golborne came down. Rivero

was summoned to explain his rebellion to the authorities above. Rivero stayed firm and refused to take the call. Gallo wondered if the commandos would have to stuff the feisty Rivero into the Phoenix, but in the end words were sufficient.

As Rivero reluctantly approached the Phoenix, the commandos grabbed his bag of souvenirs — rocks and minerals from the depths of the mine. They dumped out the rocks, handed back the empty sack and made it clear that Rivero was leaving the scene. Rivero entered the Phoenix of his own accord, and then in a final act of defiance slammed the metal mesh door shut. The miners watched in shock as Rivero slipped up and out of sight. Thanks to the luxury of seven live cameras, judicious editing and Pedro Gallo, the world saw not a single second of this center stage drama.

With Rivero and his scandal out of the way, the rescue entered the final phase. Franklin Lobos was the twenty-seventh miner to be hauled up. As the capsule climbed, he heard a deep rumbling. The crash of rock. Was the shaft compromised? How close was that one? Acoustics inside the mine were tricky. Sometimes a conversation seemed to drift down the tunnels and arrive like a whisper. Other times a vacuum appeared to suck away the words from a colleague nearby. Lobos was sure this crash was close. "It sounded like a whole level came down," he said.

At 7:20 PM, when he made it safely to the surface, Lobos was met by his daughter Carolina. He grabbed her in a tight hug. She spread her open palms across his face; for a moment they stared into each other's eyes. Carolina then handed her father a new soccer ball. He took the cue and began a dexterous display of foot juggling. Lobos's new life had begun. He would never be the same person who entered the mine ten weeks earlier. Even the smallest rituals of normalcy were now delicious.

Inside the triage hospital, an entire wall was covered with the names of the miners and the rescuers. Each time the Phoenix surfaced, a name was checked off. The celebration was beginning.

Family members crowded at bedsides to hold hands with the still-stunned miners. A cacophony of ringing cell phones, the echo of backslapping hugs and the bustle of a hundred people inside the makeshift clinic was interrupted every half hour as the latest rescued miner was wheeled in to a chorus of cheers. Doctors hugged F16 pilots.

Nurses posed with submarine commanders. Paramedics, geologists and mapmakers embraced for what was likely the last time. After months of constant teamwork and nonstop contact, the battle was nearly over.

Richard Villarroel
Juan Aguilar
Raúl Bustos
Pedro Cortés
Ariel Ticona

The list of successful rescues continued. By 9:30 PM, all but the last miner had been extricated.

Once again, the Phoenix descended deep into the mine, the prison in which thirty-three men had been trapped for over two months. At the bottom of the mine, Urzúa stepped carefully into the capsule. He took a look around and then was headed up. His mission was nearly complete.

President Piñera and what looked like dozens of aides were crowded by the rescue hole. The once strict police controls had evaporated, and spectators flooded the site. Down at Camp Hope the growing tension was about to explode. Around the world, a billion viewers stared in disbelief. What had seemed like a tragic tale of dead miners was about to be rewritten into the story of the most remarkable rescue in recent memory. Thirty-three men. Two thousand three hundred feet. Sixty-nine days. The cold facts spoke of certain death. Now the live shot of Urzúa arriving to a cheering entourage was like a fairy tale.

At Camp Hope, champagne, balloons and cheers filled the cold starry night. A community built on faith and determination had beaten the odds.

Urzúa stepped forward to shake hands with Piñera. In a tradition as old as mining itself, he symbolically passed the responsibility for the men from his command. "Mr. President," he said, "my shift is over."

As he was wheeled into the triage hospital, Urzúa, looking taciturn and serious, crossed his thick arms over his chest. His face shrouded in beard, he was the least likely of world heroes. Ten weeks earlier he had entered San José as the shift supervisor at an unknown gold and copper mine. Now he was a symbol of global goodwill. Having narrowly dodged a date with death, Urzúa was given a second chance, a new slate and a reincarnation on a scale of which most humans can only dream. While Urzúa basked in glory, the Phoenix continued to labor as each of the rescuers was slowly lifted to safety.

It was a rescue made possible by a global outpouring of generosity. Hundreds of anonymous workers turned their lives upside down to save the miners. Some built drills. Others shipped thousand-pound drill bits. Others, like Hart, guided the drills. The realm of possible solutions had been swamped by Piñera's early decision to seek help from around the globe. He later remarked that he was guided by the Russian government's stubborn refusal to seek help when the *Kursk*, a Russian submarine, sank to the ocean floor. "The Russians could have asked for technology help from England, but they didn't," said Piñera. "I personally called every president I knew and sought technical solutions."

González, the last rescuer left below, played down his bravery and said he was merely one link in the chain. He started to read a book left by one of the miners as he awaited his own exit in the Phoenix. Before leaving, he had one last desire. "I wanted to turn off the lights," he admitted. "But they wouldn't let me."

Many miners felt the same impulse to flip a switch and shut down an experience that was still too painful and recent to bear the scrutiny of full analysis.

When González was hauled up from the San José mine, the winch stopped. The noisy motors shut down, and, after ten weeks of suffering and struggle, Camp Hope overflowed with the joy of a fleeting but perfect moment.

As the last helicopter flew off to the Copiapó hospital, Pedro Gallo looked up at the dazzling desert sky. Thousands of stars winked. The heavens, for a moment, seemed closer.

"They have left a permanent record of something beautiful here."

FOURTEEN
FIRST DAYS OF FREEDOM

Wednesday, October 13 — A New Life

Inside the helicopter, Samuel Ávalos stared in disbelief: The towering machinery, the tents, buildings, roads and parking lots! While he and the seven other miners inside the helicopter had followed the rescue operation intensely below ground, the transformation of the barren mountainside into a bustling epicenter of activity was still unbelievable. The miners asked the pilot to make an extra loop over the rescue site. With the doors open, the helicopter went into a turn, banking sharply. Scanning the scene, the miners began to realize the scale of Operation San Lorenzo.

Leaving the camp, the helicopter dipped low over the desert as it followed the same road that ten weeks earlier the men had traveled on their way to the morning shift at the San José mine. Inside the chopper, the two Chilean Air Force officials escorting the miners asked for photographs and autographs and treated the men like celebrities. With their dark glasses, the men descended like movie stars from the helicopter at an army base. Crowds hung from the fence. Children climbed trees to see them. A roar of applause greeted them.

As they drove from the base to the hospital, crowds lined the road, waving flags, throwing flowers and holding up handmade signs. The miners were in shock. They had last experienced the world as down-and-out miners, anonymous to the point of being invisible. "For me it was strange. Wherever we went, people applauded," said Samuel Ávalos. "I was not very conscious of what was happening. My mind was just sorting this all out, organizing itself. I did not have much ability to process all this, to make sense of it all."

At the entrance to the Copiapó hospital, the van carrying the miners was welcomed by a heaving mass of people who had to be shoved away by an aggressive platoon of policemen. Inside the hospital, the director, Dr. Maria Cristina Menafra, welcomed the men and said it was "an honor" to provide them with medical care.

Once inside the Copiapó hospital the miners were put on the third floor. Armed policemen sealed off the entrance and even hospital personnel were strictly limited with regard to who was allowed to visit. Family members were permitted to enter but only at specific hours. The men were now subjected to a battery of blood tests, psychological questioning and X-rays.

As they delighted in the discovery of simple pleasures like a shower and a bed, the men began to comprehend the size of the media horde that had laid siege to the perimeter. For brief moments, Samuel Ávalos, who shared a room with the miner Alex Vega, could pull back the curtain at the window, poke his head out, and gawk at the phalanx of reporters toting microphones, telephoto lenses and notepads. "I would look out the window and people were everywhere. People were sleeping outside to see us."

The men lived inside a bubble. They could watch themselves on TV and listen to the nonstop commentary on what the rescue meant, on how soon they would be released, and about the alleged millions of dollars that Hollywood and TV producers were ready to throw their way.

Back at the mine, the infrastructure of Camp Hope was being dismantled by two competing squads: company employees, who were packing up their machinery and supplies, and a roaming band of rescue workers and government officials, who were grabbing souvenirs ranging from the tiny flasks used to send paloma messages to drill bits that weighed over 220 pounds. Like the Berlin Wall, Camp Hope and the rescue site were being picked and hacked to pieces by the hour.

At the rescue shaft, a round metal lid like a manhole cover was placed over the tube. Fears that curiosity seekers, tourists or adrenaline junkies might try to surreptitiously descend forced the government to keep a squad of policemen near the shaft and at key entry points to the upper levels of the hill. The mouth to the mine was virtually ignored, no shrine or permanent barrier erected.

Back at the hospital, the men basked in the richness of a breath of fresh air, an orange, a kiss and a solid roof overhead that did not threaten to cave in while they slept. The absence of dripping water was so notable that several of the men said they missed that habitual backbeat from inside the mine. The mundane routines of daily life were now deep pleasures. Ávalos described the wonder of seeing greenery, of seeing trees and the sky. "When I looked at the horizon, it felt like my brain was suddenly orienting and organizing all this information in a huge whirl of thought." Ávalos said he felt as if his life had morphed from a two-dimensional existence to three dimensions. "We appreciate life in a way that others might find difficult to understand," he added.

German tabloid reporters pushed to exploit cheap veins of gossip. Whose wife cheated on her husband? Had any of the men had homosexual experiences inside the tunnels? Who punched whom among Los 33? In their instinctual obsession with sex, drugs and scandal, the tabloids scoured the landscape — buying letters and haranguing family members in pursuit of the ultimate scandal even as serious journalists from the BBC, *El País* (Spain), the *New York Times* and other media outlets from around the world tried to get into the hospital for exclusive moments with the men.

"I hope that the avalanche of lights and camera flashes rushing toward you is a light one," wrote Hernán Rivera Letelier, a Chilean writer, as he sought to warn the miners of the media barrage headed their way. "It is true that you have survived a long season in hell, but when all is said and done, it was a hell you know. What is heading your way, companions, is a hell that you have not experienced at all: the hell of the show, the alienating hell of TV sets. I have only got one thing to say to you, my friends: grab hold of your family. Don't let them go, don't let them out of your sight, don't waste them. Hold on to them as you hung on to the capsule that brought you out. It's the only way to survive the media deluge that is raining down on you."

For the tabloids, the story was disturbingly humane. There was no corpse. No demon. No bloody climax to embellish for a brief worldwide audience. Under the cheap cover of the public's "need to know," the tabloid media pushed an agenda featuring the lowest common denominator. It was an attempt to exploit the commonly held belief that humans under extreme stress inevitably resort to barbaric behavior. As they followed their own bias, the sensationalistic media ultimately failed in its most basic task: to educate and inform.

Inside the hospital, the miners were confused. They had never considered themselves physically ill or mentally weak. With the exception of a few specific dental problems, damaged eardrums and sprained muscles, they were ready to leave. The doctors refused. A sense of protection and ownership still dominated the medical response. Few of the doctors could actually believe that the men were so healthy.

Thursday, October 14

At 8 AM, President Piñera visited the thirty-three miners in the hospital and promised to radically reform working conditions not just in the mining industry but in the transportation and fishing industries. "We can guarantee that never again will we permit that in our country you work in such insecure and inhuman conditions," said Piñera. "In the upcoming days we will announce to the nation a new agreement with workers."

Posing in hospital gowns, their sunglasses in place, the men received a challenge from Piñera: a soccer match between the presidential staff and the miners. "The team that wins will stay in La Moneda. The team that loses goes back to the mine," he joked.

The men laughed and conversed with the media-savvy president. Despite their ordeal and suffering, many of the miners were already discussing a desire to return to the mining profession. "Of course, we have to keep working; this is part of our life," said Osmán Araya. Miner Alex Vega agreed. "I want to go back. I am a miner at heart. This is something that is in your blood."

Day 3: Friday, October 15

The miners awoke anxious. Nightmares of the mine haunted their sleep. One of the miners woke up in the middle of the night and began wandering the halls, looking for the paloma. It was time for his shift and he was headed to duty. "They are dreaming about the mine," said Minister of Health Dr. Mañalich. "Others keep thinking they have duties inside the mine to complete."

Pressures mounted as families demanded to take their loved ones home and the miners pressed to be free. "We have a certain level of uneasiness because we are handing very fragile people back to their family," said Mañalich, who examined the miners as they prepared to be discharged. "It is highly unlikely that these men will go back to a normal life."

Post-traumatic stress disorder was practically guaranteed for at least some of the men. While Sepúlveda had used the crisis as a springboard to develop his latent talent for leadership, Edison Peña could not run fast or far enough to escape the pressures and trauma of the confinement. Even slightly strange sounds — like a metal cup falling on the floor — sent the men jumping. Several men slept with the lights on. Others needed sleeping pills to put their mind at ease. Psychiatrist Figueroa estimated that 15 percent of the miners might develop serious psychological problems, 15 percent would become stronger, more robust people, and the rest would be somewhere in the middle. No clear examples from history could be used for comparison. For counseling traumatized soldiers or survivors of airplane crashes a vast literature is available for psychologists to consult. The Chilean mine accident left its victims with such a unique experience of survival that few of the regular rules of mental health could be applied.

Despite the uncertainties about the men's mental stability, at 4 PM on Friday, October 15, twenty-eight miners were released from the Copiapó hospital. An elaborate ruse was developed to smuggle the miners out of the hospital, right under the eyes of the world press. While high-profile ambulances left the front gate, allegedly with miners inside, the real miners snuck out the back door. "I used to work in intelligence," said Dr. Jorge Díaz, smiling, when asked how he organized the clandestine operation.

Omar Reygadas was dressed up as a police detective so successfully that he actually waded into the media throng and began taking pictures of the journalists. Other miners switched out their clothes and sunglasses and left the hospital arm in arm with decoy women posing as their wives. Chatting and relaxed, the miners left unnoticed and were shuttled to their home or hotel.

Samuel Ávalos went to a boardinghouse where a private room with a shower and his eager wife awaited. "I attacked my wife; it had been so long. I was like a rabbit," he said. "But I could not sleep. My head was spinning. No way. My left arm twitched. My body was not relaxed. Totally tense. I was not myself. When I touched my body, I felt strange. I was not sure what I believed when I looked in the mirror. Those were not my eyes."

As the miners were released from the Copiapó hospital, they were stunned by their reception. "I didn't think I would make it back, so this reception really blows my mind," said Edison Peña. Holding back tears, Peña said, "We really had a bad time."

As media outlets lined up outside the humble homes, a new economy of information developed. Bolivian-born miner Carlos Mamani demanded a set fee per question. Other miners charged thousands of dollars, then refused to discuss any details of their entrapment. The incriminations raged as reporters felt ripped off and the miners felt justified in cashing in.

Yonni Barrios could barely reach his home. An intense media pack battled to cover his story. Though his efforts to provide health care to his thirty-two companions were a daily duty, Barrios's romantic life made far more headlines, as his wife and his lover, Susana Valenzuela, both laid claim to his heart.

Barrios chose Valenzuela as his permanent partner. When he spoke briefly to the press, he broke down in tears as he described his role as doctor: "I only did my job down there. I gave my best efforts to help my colleagues who are now my good friends."

When asked for details about the first seventeen days, Barrios refused, out of loyalty to the "pact of silence." Jimmy Sánchez, the youngest miner, however, gave an interview in which he lashed out at Urzúa as a hapless leader. "Mario Sepúlveda was the one who led us," said Sánchez in the opening volley of admissions, clarifications and declarations.

Where was Sepúlveda? The press demanded an answer. According to public statements from the hospital, Sepúlveda was fatigued and needed to rest. In private conversations, however, doctors admitted that Sepúlveda was being held — against his will — to protect him from what doctors and psychologists feared would be unbearable pressure from the media.

Sepúlveda, now entrapped yet again, this time by his rescuers, was irate. He wanted to leave the hospital.

Dr. Romagnoli came to visit and found Sepúlveda drugged and confused. Sepúlveda pleaded with Romagnoli, "Get me out of here. They are sedating me. This is an insane asylum. They are giving me shots."

Under the effects of the drugs, Sepúlveda was sleepy and nervous. "They were giving him Haldol," said Dr. Romagnoli, who described the medicine used to treat acute psychosis and schizophrenia as so strong it left the patient "knocked out."

"He was filled up with diazepam [an antianixety drug] to keep him under control. He was desperate," said Romagnoli, who decided it was time to spring Sepúlveda — even if it took a brawl. "I spoke to Iturra and told him, 'Get him out of there or it will be a major fuck-up because I will knock out a couple of cops and get sent to jail.' Happiness should not be treated as a disease."

Sepúlveda was snuck into an ambulance and smuggled to a nearby clinic, once again throwing off the media surrounding the hospital. Finally, Sepúlveda was on the verge of freedom. Katty Valdivia, the miner's wife, described her husband as perpetually hyperkinetic and buzzing with energy. "They don't understand Mario — he is like this."

Saturday, October 16

Thirty-one men had been released from Copiapó by Saturday, October 16, but Sepúlveda and Zamora were still confined. Zamora had a badly infected tooth. At 10 AM, Sepúlveda was released, and he immediately headed to a rented apartment for a much delayed birthday celebration with his family. He had spent his fortieth birthday, on October 3, trapped underground. Now he was ready to celebrate with Katty and his children.

The hyperkinetic Sepúlveda sat little during the birthday meal. He bounded into a back room and, like a child discovering lost toys, began to open the sealed packages he had sent up from the mine. He hauled a crate of crudely wrapped tubes the size of baseball bats into the living room and jammed the blade of his pocketknife into the thick plastic of one of the packages. With desperation, he began to saw open the tube and then pulled out the contents: plastic bottles filled with minerals and crystals from inside the mine. "These are the remnants of the explosion, when we were first trapped and no one could hear us," explained Sepúlveda. "This is a symbol of our hope and our attempts to escape. I only offer these to people who are important to me."

Sepúlveda then began to pull out the letters he had received below. As he read them, his face changed. The smile disappeared. Tears rolled down his face. His choked up as he tried to explain the flashbacks and hard memories flooding his mind. Then he announced that he wanted to travel. Now. To the one place he dreamed about most during his confinement: the beach.

During the drive to the beach, Sepúlveda spoke like a man recently released from jail. Everything attracted his attention, from the sounds of traffic to the ease with which he could buy food, choose a soft drink and move freely. Sepúlveda said, "Now I value everything." He picked up two empty plastic water bottles from the floor of the car. "Look at these; with these two you can make a shower. One to soap up and then you rinse with the other."

The beach near Caldera was abandoned. The sun hid behind a bank of gray clouds. A warm breeze wafted from the ocean and flocks of seagulls hunted for scraps at the water's edge. Sepúlveda began to play soccer with his son. Then he stopped to appreciate the moment. "You know what I always dreamed of when I was trapped? This was my most precious dream, to bathe at the beach!"

As he spoke, the sunlight cut through a hole in the clouds and angled shafts of golden light reflected on a patch of ocean. "That is the divine light, the light of hope," said Sepúlveda. "When that first paloma came down to us, I pointed to the hole and told my companions, 'That is the light and the door to hope. And above, my friends, is paradise.'

"I want the world to learn from us. To learn how to live. We all have our good and bad sides; we need to learn how to cultivate our good sides," said Sepúlveda, who lived firsthand the fragility of life. "Your life can be over in two minutes. What good is all the money if you are not alive? No, look at me, I am happy. Two months without a penny and I am happy." Pointing to the waves and sky, he said, "This is life."

Sepúlveda flipped a soccer ball, ran on the beach with his son, chased seagulls and then pulled off his shirt, kicked off his shoes, dropped his shorts and, buck naked, arms outstretched, ran into the surf. As the waves lapped around his ankles his family applauded.

Mario Sepúlveda, the leader of Los 33, bathed and frolicked like a child.

Sunday, October 17

In an effort to close the cycle of trauma and make peace with their experience, twelve of the miners went back to the San José mine four days after their rescue. A Thanksgiving mass was planned at the remnants of Camp Hope and religious, political and rescue leaders were gathered to finally put an end to the odyssey.

Under a tent guarded by police officers, the scene quickly degenerated into a screaming fight between the police and a group of workers from the San José mine who were not allowed to enter the religious service. Under the guidelines developed by the government, only Los 33 were invited to the service. Fellow workers from the San José mine were irate. They had suffered as well, many volunteering for weeks to help rescue their fellow miners. This was their work site; their sweat and suffering were now part of this mountainside and they felt they were being treated like interlopers. Shoving matches with policemen broke out. Protesters from the mine began to denounce the failure of the mine owners to pay back salaries.

■ ■ ■ ■

Security guards also denied entry to representatives of the miners' union who arrived to both commemorate the successful rescue mission and protest the back wages owed workers. The mine owners were invisible as they presumably plotted a defense against the anticipated wave of lawsuits, a process likely to bury them in legal proceedings for years. Would they be sentenced to jail? Forced to pay huge time? The wheels of Chilean justice would take time to reveal the outcome.

"They want to pay us in installments over eleven months," said Evelyn Olmos, a union spokeswoman as she criticized the owners of the San José mine. "We need the money now."

"When nobody thought that our companions were alive, we all came here. All the workers — we knew they were alive and we brought our solidarity and faith," complained Javier Castillo, a local union leader. "And now that they are alive and it is back to business as usual, they shut off the mass for a few of us. This is painful."

Castillo had long fought for the mine to be closed. For nearly a decade he had watched as a seemingly endless series of accidents maimed and killed workers. The casualty list was as much a part of the mine, it seemed, as the steady stream of trucks hauling valuable gold and copper from the bowels of the mountain. As further details about the safety conditions in the mine came to light, a backlash built against the mine owners.

Manuel González, the first rescuer to descend in the capsule, was appalled by the conditions inside San José.

"It didn't even have basic elements," he told TVN, the state television. "I was down there for twenty-five hours in temperatures of forty degrees Celsius [100 degrees Fahrenheit] . . . and there was almost one hundred percent humidity. I imagine those first seventeen days when they did not know anything . . . it must have been terrible."

After attending the mass and waiting for the media to disperse a bit, Samuel Ávalos disguised himself and began exploring Camp Hope. It was a foreign world. In the months before the collapse, he had driven daily through this same area, past mounds of sterile rock. Now every corner was packed with cables, motor homes and signs of life. Ávalos was given permission to visit the shaft that had brought him to safety. "I found the hole very small," he said. "I still can't explain how I came out of that hole. If you ask me?

I can't understand — without a doubt it was a rebirth."

Then he began insulting the mountain. "This was a mine that was vicious. When you insulted it, it would throw rocks at you; it was a living being. . . . So I pissed on that fucking mine and as I did I insulted it and called it names." But even as he sought vengeance, Ávalos maintained a deep respect for the mine. "If the mine wants to kill me, it will, even if I am out here. It has a power."

Another group of rescued miners left the mass and headed to the mouth of the mine. Standing outside, they stared in at the yawning mouth. Then they picked up a handful of rocks and, screaming insults, hurled the stones into the hole. For a moment, they had won; the mine had no answer.

EPILOGUE
THE TRIUMPH OF HOPE

It was a combination of fate and last-minute decisions that led the thirty-three miners to enter the San José mine on August 5, 2010.

Mario Sepúlveda missed the bus to work. He hitchhiked on a lonely road that fateful morning, arriving hours late. Samuel Ávalos was not even a miner. He sold pirated compact discs on the streets of a small Chilean city; a relative took him to San José for a new opportunity. Carlos Mamani did not even have a contract to work inside San José. He was moonlighting to make a little extra cash to support his newborn daughter, Emili.

Every time a new shift began at the San José mine, the men entered a world that was known to charge tribute, a mine known as a vengeful spirit that never hesitated to show its wrath by raining a shower of rocks upon the working men. Los 33 were not regular guys. They were victims even before the mine collapsed. It took a string of bad luck, hardened circumstances and blazing bravery to even consider working at the San José mine.

Among Los 33, accidents were part of the daily gamble. If a man could finish a twelve-hour shift without being crushed, he pocketed another $75. If he managed to survive a full week, he earned $525. With extra days and overtime, some of the men managed to make $2,000 a month. The San José mine paid roughly 30 percent more than mines of a similar size in the region, a practice not unlike military forces and diplomatic corps that allocate extra pay for working in a war zone.

When the mountain collapsed on August 5, men should have been killed. At any other time of the day or night the massive cave-in would have crushed and forever buried at least some of the scattered crews inside the labyrinthine mine. But the mountain cracked at lunchtime — just as the men were retreating to the refuge for lunch, stowing their tools or preparing to ride the transport truck up to the blazing sun, fresh air and food. Once they were locked behind a wall of rock the size of a skyscraper, the miners had virtually no food and no way to escape. It was estimated that clearing the tunnel, digging through the rock would take a full year.

As they slowly starved to death for seventeen days, many of the miners cursed their luck — "If only . . ." "Had I just . . ." "Why me?"

The slow death gave them more than sufficient time to examine their lives, to take stock of their accomplishments, failures and families. The sum was not flattering. Many of the men had squandered their earnings on cheap thrills, leaving wives or girlfriends and children to fend for themselves. Others had succumbed to alcoholism and drug addiction. Generosity and altruism were not notable traits among the group.

The work routines inside San José were hardly conducive to introspection or self-improvement. Daily dangers were so prevalent that it was hardly surprising that after seven days of dodging danger the men would feel justified in blowing their salary on cheap booze, secret lovers and other equally disposable placebos.

Then a miracle happened. Instead of succumbing to animal instincts and the behavioral meltdown so embodied in the novel *Lord of the Flies,* the miners grasped the essence of the human spirit, and they never let go.

Instead of fighting over a can of tuna, they divided the meager contents into thimble-sized portions. A single can of peaches became a communal feast. Rather than let brute force rule, they instituted a daily meeting where key decisions were discussed, debated and then put to a vote. "Humor and democracy," said Luis Urzúa when asked how he had helped guide the leadership during ten weeks underground. "We were thirty-three, so sixteen plus one was a majority."

In the first hours after the accident, family members of the trapped men rushed to the scene, built shrines, and implored politicians never to give up as they deliberately turned their collective back on logic and probabilities. In their hearts they held fast to a distinct belief: of course the men were alive. The only doubt was how long their loved ones could keep fragile grasp of life.

For ten weeks, thirty-three men united and battled together. Buried 2,300 feet deep, inside a mountain, they built a communal spirit to survive. Even as their families rallied and rescuers launched multiple plans to save them, the Chilean government drew up funeral plans and a design for a white cross to adorn the hillside in their memory. Statistics provided to President Piñera suggested that there was only a 2 percent chance that *any* of the miners had survived.

Faith and technology were ultimately united to literally move a mountain. Increasingly the families, the miners, the rescue workers and the world media demonstrated the ability to work for the common good.

The final cost of the rescue was estimated to be in the range of $20 million — roughly $600,000 per miner. Not only was the final tally for the rescue rarely questioned, in many cases the bills never arrived. Mining companies including Precision Drilling, Minera Santa Fe, Center Rock, Anglo American, Geotec, Codelco, Collahuasi and dozens more simply paid out of their own pocket.

As the rescue effort labored on, donations poured in from Japan, Canada, Brazil, Germany, South Africa and the United States. United Parcel Service shipped 26,000 pounds of drilling equipment for free. Oakley sent a small box with thirty-five pairs of sunglasses. Teams of machinists at Center Rock Inc. in Pennsylvania worked overtime to design a new drill bit. For every trapped miner, an estimated thirty to fifty people worked full-time to assist in the rescue. At lunchtime, the mess hall at Camp Hope sounded like the United Nations — Korean journalists, Brazilian oil workers, NASA doctors, Chilean firefighters, Canadian roughnecks and, from Colorado, the towering Jeff Hart, the world's best driller.

When queried about the lessons of his entrapment, Samuel Ávalos, said, "We are as fragile as a second. In the least expected moment, it is all over. Live and enjoy the now. The instant. The moment. Don't make too many plans. Understand that your problems are so much smaller than what we lived. . . . Always have the capability to overcome, to help others."

How did a ragtag band of desperate miners and their families become a showcase of tenderness and emotional intelligence? Few of these men were well educated, successful in their career or able to spend "quality time" with their families. They were hardened men, survivors who labored in anonymous corners of a dark cave where few other humans could last a single shift.

Colleagues were killed. Colleagues were maimed. New recruits rushed to fill the open slots. For these workers — and they exist in every corner of the world — the concept of a just universe or meritocracy was as foreign as the procedures for boarding an airplane or applying for a passport. Yet they became an example to the world, a symbol of survival. A brief reminder that like evil, good exists. And a reminder that in an ever more connected world, a single event has the power to unite us.

When a group of zealots launched the attacks on the World Trade Center in 2001, the world was instantly torn apart. For the worst of reasons, divisions suddenly overshadowed understanding. Racism. Tribalism. Us versus them and "shock and awe" erased a nascent global consciousness. Then the French newspaper *Le Monde* published the famous headline "We Are All Americans." This was not a celebratory moment. Indeed it was a declaration of defeat, an admission that it was now time to confront the most brutal tactics with even more brute force. An era of terrorism and torture arrived. Guantánamo became a symbol of the new Dark Age.

The Chilean mine rescue was the Anti-9/11, an event that showcased human charity, brotherhood and the concept of a Global Village built on altruism. The world media fixation with the Chilean mining story was an aberration from the normal flow of war news, massacre updates and extreme weather. Was it a flash in the pan? Or was it a brief glimpse into the vast reservoir of goodwill that can always be summoned for a worldwide cause.

The global embrace of the Chilean miners had as much to do with the state of the planet as it did the fate of the trapped men. Every year, thousands of miners are trapped and die. Hundreds more are rescued. The world's press has no shortage of global good-news stories. Heroes abound if reporters and editors take the time to search. After nearly a decade of what analysts call "the Age of Terror," by August 2010 the world seemed starved of hope, but the bravery of thirty-three men and a band of generous and tenacious rescue workers brought the world together. At least for a moment, we could say, "We are all Chilean."

AUTHOR'S NOTE

A Word on Translation

Chilean Spanish is notoriously rich in slang. Miner lingo is notoriously rich in obscenities. The combination of those two realities made a literal translation virtually impossible. In many cases the Spanish used by miners and rescue workers, families and politicians has been translated to preserve the essence and meaning. Many obscenities have been simply removed not because of any particular offense, but due to the simple fact that they do not make sense in other languages. The author and publisher have maintained the intent of the words but allowed for a more coherent translation style. Given the use of multiple translators, there are likely to be minor differences of opinion in how best to present the rich texture of Chilean Spanish to a global audience.

A Word on Dates and Hours

This book is based on interviews with approximately 120 different participants in the rescue, including a majority of the miners, President Piñera and leading designers and participants. Due to the extraordinary nature of their underground confinement and the monotony of confinement, the miners were not always able to confirm the exact time and date of certain events. With no daylight or darkness to mark the passage of time, such confusion is understandable.

The author has sought to make sense of this confusion and understands it well as he personally went for one eight-day stretch without changing clothes, showering or even taking off his boots. Fatigue and exhaustion were in abundant supply for the last twenty days of the rescue. The author wishes to stress that despite repeated efforts to clarify certain sequences, discrepancies continue to exist among the very participants. Such is the nature of dramatic events.

Exclusive Access

Many of the scenes and interviews in this book were not available to the thousands of journalists at Camp Hope. Early in the rescue effort I reported from behind the police lines like the other reporters. As I realized the scope and drama of the operation, I asked the Asociación Chilena de Seguridad (ACHS), the insurance company in charge of much of the rescue operation, for permission to document their remarkable rescue efforts. I was covering the drama for numerous media including the *Washington Post* and *The Guardian*. The ACHS immediately agreed and provided me with a half-day tour of the rescue operation. They then told me the tour was over.

I asked to stay on and continue to report. In that case, I was told, I would need a Rescue Team credential. I filled out the forms, wrote that I was a writer and was given pass No. 204. For much of the ensuing six weeks I was able to roam the front rows of the rescue as I reported, recorded and filmed. At no time did I suggest that I was on any other mission than full-time reporter.

The Oakley Sunglasses

In accordance with full disclosure, I am proud to say that I was responsible for the miners receiving Oakley sunglasses. During a planning meeting between Codelco and the Chilean Navy early in September, it became apparent that the miners needed high-quality eyewear upon leaving the mine. Given the overwhelming logistics of planning the rescue, the officials were swamped with tasks and a bit lost on what to do about sunglasses. Seven years earlier, I had met Erik Poston, an Oakley representative. I still had his business card. I wrote Oakley an email suggesting they send thirty-five pairs of sunglasses (two spares) to the Chilean rescue team. They complied and the rest is history.

ACKNOWLEDGMENTS

In writing this book in the immediate aftermath of the mining drama, I found the challenges were numerous, the sacrifices many. First I would like to thank my wife, Toty Garfe, for accepting my months-long disappearing act. And to my daughters Kimberly and Amy, sorry to have missed your birthday. Zoe, your baptism photos are great; it would have been better to have been there. Susan, congratulations on winning so many high-diving medals; I saw the video. Maciel, how did you grow half a foot in two months? Francisca, my first daughter, your loyalty to your globe-trotting dad is appreciated.

To Annabel Merullo, Caroline Michel, Juliet Mushens, Alexandra Cliff and the team at PFD, you were the first to see the potential of this book and guided it through the Frankfurt Book Fair and to higher ground. I am forever grateful. George Lucas of Inkwell Management, you guided this book through the jungles of U.S. publishing and landed me at Putnam, where Marysue Rucci and Marilyn Ducksworth were instrumental in turning this book into a beautifully designed, finely edited and nationally known work. Both Diana Lulek and Michelle Malonzo at Putnam dropped their schedules to answer my many questions about publishing my first book. At Putnam I would also like to thank the managing editorial team of Meredith Dros and Lisa D'Agostino, who were busier than air traffic control at JFK, keeping all the pieces of this project in sync. And to Putnam President Ivan Held for his backing of my first book: I appreciate your unwavering support. Finally, I want to thank Art Director Claire Vaccaro, who patiently figured out the maps and photos, and Copy Chief Linda Rosenberg, who worked through the holidays. To Bill Scott-Kerr and Simon Thorogood at Transworld Publishing in London, your early support for this project was key to making it happen. Bob Bookman and the team at CAA for everlasting optimism inside the madness that is Hollywood movie production. To Colin Baden, Diane Thibert, and Rachel Mooers at Oakley for a generous contribution to the miners. Thanks to Martín Fruns, Alejandro

Piño and the entire medical and psychological staff at the ACHS for their professional treatment and constant help throughout this project. A special mention of psychologist Alberto Iturra, who may have had the hardest job of all — keeping the miners united underground. Alberto, the miners may not have realized how hard your job was, but the rest of us did!

My father, Tom Franklin, used to fact-check and copyedit my third-grade homework — see, it all paid off! My older sister Sarah, who is a tremendous writer and even greater inspiration, cleared the way for my career. My younger brother Christopher, who silently creates parks and recreation areas, your legacy is already here. To my mother, Susan, who is watching from above, you not only delivered me to this world but imbued me with a spirit of survival and endurance.

To my colleagues, including Dean Kuipers, *L.A. Times,* early investor in my self-confidence. Sam Logan, of SouthernPulse, a visionary journalist. Denise Witzig, Brown University, a key mentor. John Kifner, *New York Times,* early mentor. Hunter S. Thompson, no explanation needed! Michael Smith, Bloomberg, best investigative reporter I know. Jorge Molina, *El Mostrador,* Chile's finest reporter. Pablo Iturbe and Tim Delhaes, Tigabytes, close conspirators and fellow dreamers. Rory Carroll, Martin Hodgson, David Munk, Marx Rice Oxley and the entire foreign desk at *The Guardian* for putting up with my random acts of generosity. Tiffany Harness, Doug Jehl Griff Witte and Juan Forero at the *Washington Post,* who proved that great editors still exist. Guillermo Galdos, Discovery Channel, for hosting me at the San José mine and always providing inspiration. Lonzo Cook and Karl Penhaul of CNN for good humor and great dinners. Amaro Gómez-Pablos Benavides of Televisión Nacional de Chile, for loyalty and laughs. Francisco Peregil from *El Pais,* an example that great journalism can also be collaborative. Bert Rudman, John Quinones, Joe Goldman and the whole ABC News team, who took me under their wing at Camp Hope. Carlos Pedroza and Manuel Martinez of *Esquire Mexico* for their long-term loyalty and eye for real news, Miguel Soffia for proving that the next generation is going to make us old-timers feel slow and lazy! And finally to James Bandler, my eternal co-conspirator, master of the understatement and the world scoop and the reason I went to Chile twenty-one years ago.

And to the thirty-three miners — each of

whom took time to speak with me and provide information for the book. In particular I would like to thank Mario Sepúlveda, Raúl Bustos, Alex Vega, Juan Illanes and Samuel Ávalos.

Finally my business partner Morten Anderson, for his patience during my unexpected sabbatical. Assistants Gemma Dunn, Lucia Bird and Ellen Jones were infinitely patient in transcribing Chilean miner interviews from an almost unintelligible slang into perfect English.

ABOUT THE AUTHOR

Jonathan Franklin is an award-winning journalist who reported from the front line of the Chilean mine disaster for the *Guardian*, the *Washington Post* and the *Sydney Morning Herald.* As the only print journalist with front row access to the rescue efforts, he was allowed to attend planning meetings, private conversations with the miners, and to have practically unlimited time with the lead doctor and the lead psychologist. He was also the first journalist to secure an interview with the leader of the miners, foreman Luis Urzua. He has lived in Chile for 16 years.

The employees of Thorndike Press hope you have enjoyed this Large Print book. All our Thorndike, Wheeler, and Kennebec Large Print titles are designed for easy reading, and all our books are made to last. Other Thorndike Press Large Print books are available at your library, through selected bookstores, or directly from us.

For information about titles, please call:

(800) 223-1244

or visit our Web site at:

http://gale.cengage.com/thorndike

To share your comments, please write:

Publisher
Thorndike Press
295 Kennedy Memorial Drive
Waterville, ME 04901